Operating Employee Associations:

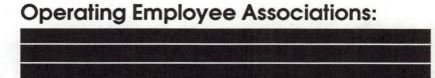

Providing Employee Services and Recreation Programs

Brad Wesner
University of Illinois

Sagamore
Publishing
a division of
Management Learning Laboratories

ISBN:0-915611-22-8

This is the first in a series of books on Employee Associations produced by Sagamore Publishing in cooperation with the National Employee Services and Recreation Association. Other books forthcoming in this series include: *Adult Programming for Employee Services*, *Volunteers in Employee Service Organizations*, and *Revenue Generation and Management*.

ISBN: 0-915611-22-8
Library of Congress Catalog Card Number: 89-060960

This book is dedicated to all of those who will not accept mediocrity as a way of life.

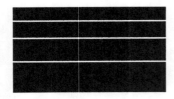

Acknowledgments

Employee associations - what are they, why do they exist, how did they come into being, and how does one operate their facilities? Employees, employers, administrators, and even the general public have asked these questions, and this book attempts to answer them and serve as a reference book for administrators of employee associations.

Numerous reference books and practitioners were consulted in the writing of this book. Special thanks goes to Steve Coffey of the Lockheed Employee Recreation Association, Bob Crunstedt of Honeywell, Inc., Jenny Izzie of Chemical Abstracts Service, Randy Schools of National Institute of Health, Dennis Mullen of Pratt & Whitney Aircraft Club, and Sue Potter of Nationwide Insurance Company - the National Employee Services and Recreation Association (NESRA) Book Review Committee, and to Patrick Stinson, director of NESRA. A personal thanks goes to Dr. Joseph Bannon, Professor in the Leisure Studies Department at the University of Illinois, Urbana-Champaign, and his Sagamore Publishing staff, Peter Bannon, Susan Williams, and Michelle Dressen for their time and input. Also thanks to Dr. Tony Zito, Dr. Bill Brattain, and Mr. Jim Miner who introduced me to facility management at Western Illinois University.

This book is a compilation of many people's thoughts, many of which have never been written, as well as my own. Although this book is a stepping stone in the development of the employee association, it is by no means the end product. The employee association is dynamic and growing, and the more we know about it the better it can work for all of us.

Brad Wesner
June 1989

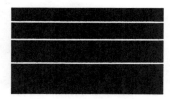

Foreward

In 1979, NESRA published a college textbook entitled *An Introduction to Industrial Recreation: Employee Services and Activities*. With an obvious need for an update, the NESRA Research and Professional Development Committee studied the alternatives that would best meet the membership interests. The conclusion was to publish practical, informational books on topics of significant importance that would serve as handy reference material for new members in the field as well as experienced members.

NESRA is proud to present the first publication in this four-book series, *Operating Employee Associations-Providing Employee Services and Recreation Programs*.

Throughout the history of our field, employee associations have played an integral role in the administration of employee services and recreation programs. Employee associations weave a common thread through the diversity of our membership–large companies, small companies, companies with facilities, without facilities, urban and rural locations, government organizations, private corporations, large budgets and small budgets, highly structured and informally structured, professional staff or managed by volunteers–each have found employee associations to be beneficial in managing programs.

This book offers and examines the essentials needed to run a successful employee association. It focuses on the history of employee associations, tasks that should be dealt with, the management of people and offers an extensive number of samples for employee association administration. The remaining books in this series will be published in the latter part of 1989 and by midyear 1990. The subjects that will be addressed are: working with volunteers, adult programming and financial management.

Employee services and recreation has taken a vital role in meeting the needs of today's workforce and will be even more of a factor in tomorrow's workforce. These books will assist the practitioner by providing state-of-the-art information.

On behalf of NESRA, I wish to thank the Research and Professional Development Committee chaired by Vice President Sue Potter of Nationwide Insurance Company; the book review committee consisting of Steve Coffey of Lockheed Employee Recreation Association, Bob Crunstedt of Honeywell, Inc., Jenny Izzie of Chemical Abstracts Service, Dennis Mullen of Pratt & Whitney Aircraft Club and Randy Schools of National Institutes of Health; the NESRA Board of Directors; and the NESRA Education and Research Foundation who financially supported this project.

Patrick B. Stinson
NESRA Executive Director
June 1989

 *Preface*

The purpose of this book is to describe the operations of an employee association. That may sound like an impossible task since no two employee associations are exactly alike. Some employee associations consist of a few clubs while others have employee stores, snack bars, and fitness facilities as well. Some rely primarily on members for funding, while others depend upon the company. Some are staffed almost entirely by volunteers, while others have several paid personnel. Although there is great diversity among employee associations, all have enough in common that they can be examined collectively.

The first step toward operating an employee association has little to do with operations; one must learn the history and philosophy of employee associations. By understanding these, one can attempt to grasp the present state of employee associations and also plan for the future. The history of employee associations and beliefs about the mission and role they play all greatly affect the day-to-day operations of the employee association.

Once the history and philosophy of the employee association is understood, the prongs of administration– task management and personnel management–can be explored. Task management includes forming a club, establishing policies, and developing a plan. Personnel management includes working with programmers, working with outside people who have a stake in the agency, and communicating one-to-one with co-workers.

Although the focus of an employee association is on people, and coordinating the facilities is merely one of numerous ways an administrator can help employees to feel better about themselves, all administrators should have a grasp on the concepts of facility management. Although facilities are secondary, understanding how they can be wisely utilized greatly increases the odds of having a successful employee association.

This book is the first in a series of books sponsored by the National Employee Services and Recreation Association, and is intended to lay the foundation upon which the others will build. Because it is a broad overview, many topics are briefly touched on, though they deserve far more depth. This depth is provided in the other books in this series: *Adult Programming for Employee Services, Volunteers in Employee Service Organizations* , and *Revenue Generation and Management* .

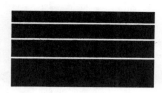

Contents

1

The History of Employee Associations

Today's employee associations are one of the major organizational forms that have evolved for the provision of employee services and recreation programs. Although the National Employee Services and Recreation Association (NESRA) did not form until 1982, the history of employee services can be traced back hundreds of years.

PREINDUSTRIAL ROOTS

Employee services have been a part of civilized society from the beginning. Under the Greek and Roman empires, many property owners provided their workers and slaves with special benefits such as discounts on the products that they grew. Under the feudal system that followed the fall of the Roman empire, peasants working for the king often received special benefits beyond their agreed-upon wage, such as sporting events and news from a town crier. Although these three examples are a far cry from the employee stores, discounted tickets to professional sporting events, and newsletters that exist today, they did establish the beginnings that eventually became the employee services we know today.

THE DARK AGES OF EMPLOYEE SERVICES

Ironically, for over one hundred years, the more industrialized a nation became, the less employee services were offered. At the beginning of the industrial revolution, employers began to see everything — including employees — as parts of a machine. Each part was interchangeable, therefore little or no attention was paid to the employee as an individual. An employee's pay was generally as low as the market would allow, and included no benefits of any kind.

The industrial revolution began in Western Europe in the latter half of the eighteenth century. Although the industrial revolution developed quickly in Europe, it was slow in coming to North America for a number of reasons, one of which was the diplomatic relationship between the United States and Europe. Colonies had never revolted from a mother country prior to the Revolutionary War, and neither France nor England expected the new nation to survive, even if it did win independence. When the American colonies revolted from England in 1776, England ordered its citizens not to give or sell processes, tools, machines, or other forms of technology to the United States, and this embargo seriously curtailed industrial development in this country.

Ironically, it was an industrial invention rather than the technology embargo that probably hindered industrial expansion more in the United States. Cotton farming in the United States was not extremely profitable prior to the invention of the cotton gin, and the agricultural economy of the South was on the verge of becoming industrialized. However, because of the invention of the cotton gin and the huge profit to be made that followed it, the South focused on the development of its cotton resource instead of developing industry. Meanwhile, the cotton gin furthered Northern industrialization which helped it to expand far more rapidly than it otherwise would have grown. This invention also helped establish an aristocracy of employers in both the North and South, who sometimes snubbed the average worker, regardless of whether the worker was black or white.

Little attention was paid to the worker at the dawn of the industrial revolution. However, when supervisors sought to increase factory efficiency, people such as Frederick Taylor experimented with ways to maximize productivity. Numerous variables were analyzed, including the worker. Although Frederick Taylor was a strong believer in scientific management and top-down hierarchies of authority, he may have set the foundation for humanistic management and employee services when he noted that people work better in certain conditions than in others.

MODERN ROOTS

With the conclusion of the American Civil War came the destruction of the agricultural economy of the South. The war brought financial ruin for some people, while others were able to amass a fortune. Those newly wealthy found plenty of opportunities to invest in both the North and South, and this investing led to the era of big business in the United States.

At first these employers tried to bully workers as their predecessors had done, but the democratic values of American society led to the formation of labor unions, so the employees could receive both a voice in policy and a share of the profits. Soon employees found themselves bargaining for things past generations had not even considered possible — paid afternoons off, paid vacations, and pension plans.

Europe had also changed dramatically. For instance, the German tribes had united into one German nation, and in the 1880s the national government had ordered that companies provide such things as sick leave, work-injury benefits, maternity leave, and unemployment compensation.

In fact, Western society in general had changed. The work ethic was fading rapidly, and the value of leisure was being recognized. No longer was one's identity tied only to one's work. As the benefits of leisure gradually became acknowledged, people began to expect more from their free time than just rest before going back to work. When employers began to see how an employee's wise use of leisure time could greatly enhance job performance, companies began to look for ways to supplement their employees' leisure experiences.

THE EMPLOYEE ASSOCIATION TODAY

The current status of employee benefits can be attributed to four primary sources: the federal government, labor unions, companies, and employee associations. Using the United States as a case study, let's look briefly at each.

Federal Government

By passing legislation, the federal government has made sure that all employees of large companies are entitled to at least minimal benefits. In 1911, Congress passed the state worker's compensation law; in 1935 it mandated unemployment benefits; and in 1978 it outlawed mandatory retirement before age 70 in the private sector. In addition to the legislative branch, the judicial and executive branches of the federal government have also played a major role in determining the benefits package. Both the courts and governmental organizations such as the National Labor Relations Board, have ruled about what is and is not bargainable between unions and companies. By making laws and interpreting them, the federal government has greatly shaped the current standard benefits package offered by employers.

Labor Unions

Labor unions serve as the voice of the workers. Because presidents of large companies cannot meet with each person individually to determine benefits and wages, representatives of the employees do the negotiating for everyone. Labor unions are more likely than the federal government to break new ground in bargaining for benefits — breakthroughs have included extended sabbatical vacations, dental insurance, bonuses, and unemployment benefits. Not only do labor unions seek to negotiate new benefits, they also strive to protect the current ones.

Companies

Although the motivation is generally economical rather than humanitarian, companies have often taken the initiative in creating employee benefits. Companies in the twentieth century have begun to offer such things as paid breaks, shorter work weeks, and company picnics, with little or no pressure from the government, the union, or the employee association.

Employee Associations

Employee associations were first thought of as primarily employee recreation associations. The founders of these associations were often volunteers who had agreed to help coordinate the annual company picnic. The first activities offered by these fledgling organizations were company picnics, sports teams, and children's Christmas parties. Soon, the job of recreation director developed into a formal position, and directors from different companies began to network through informal gatherings and by joining the National Recreation Association, which is now the National Recreation and Parks Association. The National Recreation Association, however, was primarily concerned with parks, so in 1939, seventeen industrial leaders formed a new organization, the Recreation Association for American Industry, which was headquartered in Chicago, Illinois. During the 1940s the name changed twice, to the Industrial Recreation Association (IRA) and finally to the National Industrial Recreation Association (NIRA).

A few decades later, leaders of employee associations found themselves planning far more than traditional recreation programs such as picnics and softball games – they were planning fitness programs, tours to foreign countries, social activities, recreational facilities, and employee stores. Because of this widening emphasis, NIRA changed its name to the National Employee Services and Recreation Association (NESRA) in 1982.

WHAT IS NESRA?

In 1982, NIRA recognized that employee services serve the company as well as recreation and gave birth to NESRA. Accompanying this name change have been efforts to professionalize the field of employee services and recreation, and to establish it as a viable part of human resources management. Today, NESRA, the oldest association in the human resources management field, represents over 22 million employees.

NESRA acts as a communication center providing information regarding current trends in employee services, recreation, and health and fitness. NESRA also assists in developing employee services programs — fitness, sports, employee assistance, travel, education, and discount programs — as a benefit to business, industry, and government organizations. Included in the many educational and professional services that NESRA offers to its members is *Employee Services Management*, a professional journal containing the latest concepts, trends, and how-to information. NESRA also keeps its members informed with *Keynotes*, a monthly newsletter covering NESRA news and current events affecting the field of employee services. In addition, NESRA offers an annual conference and exhibit where educational sessions and seminars are conducted, and exhibitors display the latest in employee programming products and services.

The primary responsibility of NESRA is to enhance the professionalism of employee services and recreation managers, to strengthen employer-employee relations, to generate a spirit of cooperation and goodwill among employees, and to enrich the lives of individual employees and their families. In fulfilling its mission, NESRA is constantly working to maintain high standards of performance in all phases of employee program management. In addition, NESRA is continually promoting the development of effective employee services and recreation programs and creating new educational programs and seminars. More information about NESRA and how to join is available from its headquarters: NESRA, 2400 S. Downing Avenue, Westchester, IL, 60153.

FUTURE OF EMPLOYEE SERVICES

With the government assuming a wait-and-see approach, the union's presence declining, and companies assuming a hands-off attitude, employee associations have moved to the leadership forefront. Programs within the employee associations have become even more diverse since 1982, and show promise of continuing that trend. During the past decade, the United States has become more of an information society, and if this trend continues, employees will need both informational programs to keep them updated, and exercise programs to keep them physically fit. These programs will be necessary because employees get little information or exercise standing on an assembly line or sitting behind a desk. Other trends that indicate more diversity of programs in the future, are society's desire for more dependent care services, the growth of industrial conglomerates; the growing emphasis on the individual; the formation of more part-time jobs; the need for flexible work schedules; and the increasing concern for the elderly.

Although many future challenges lie ahead, employee services—particularly those offered by the employee association—have much growth potential. As decisions become more complicated, training will be needed for those who lead employee associations. Despite becoming more formal to cope with day-to-day activities, the employee association remains exactly what it has always been—an association of employees—because despite a changing world, the values of the employee association have not changed.

SUMMARY

This chapter highlighted the development of the employee association. To understand the employee associations of today, it is important to understand the heritage of the employee association. The heritage is a rich one, and, as the present becomes the past, it becomes richer each day.

Philosophy Underlying the Employee Association

Although employee associations may appear considerably different on the surface because of differing amounts of programming and facilities, all employee associations have similar philosophical make ups, and it is this–not programs or facilities–that truly defines an employee association.

MISSION OF AN EMPLOYEE ASSOCIATION

Each employee association is founded on the basic belief that meeting the employee's physical, emotional, mental, and social needs will lead to higher morale, less job turnover, reduced absenteeism, and fewer on-the-job accidents. Research findings do tend to suggest that an employee association contributes to these things.

Services of the employee association benefit both the employee and the employer. However, employees and employers often have different motives for desiring the services. Employers frequently cite motives such as

- being truly concerned for the employee.
- developing employee pride in the company.
- improving employee morale.
- motivating employees.
- improving the company image in the eyes of employees.
- attracting new quality prospective employees.
- increasing employee satisfaction.
- reducing absenteeism.
- decreasing turnover rates.

- providing more security for employees.
- taking advantage of group rates.
- remaining competitive with other companies.
- improving safety by having more alert and better informed workers.
- improving relations with the community.
- promoting social interaction between different levels of employees (i.e., management, supervision, clerical, janitorial, and production).

Because employees are also diverse, a wide range of motives for desiring employee services exists. These motives include

- saving money.
- meeting new friends.
- learning new skills.
- staying in shape.
- socializing.
- getting ahead in one's career.
- simply having fun.

Like employers, some employees may have several motives, while others are focused almost exclusively upon one of them. Although personal reasons for participating in the employee association vary, employees generally agree that the employee association is democracy in action, because it is employees doing what they want to do.

Although employers and employees usually have different motives for desiring employee services, they can generally agree on specific services to offer. These services may include sports programs, fitness instruction, travel plans, education seminars, discount programs, and pre-retirement plans. To find out which programs would be best in a particular situation, the person coordinating employee services may ask all employees to complete a survey instrument, as well as talk to the head of personnel, the union chief, and a focus group of employees. These methods are also useful for deciding which branch to offer within a program. For instance, travel programs may be broken into fishing, hiking, tours, and cruises.

EMPLOYEE SERVICES COMPARED
TO OTHER BENEFITS

Employee services are only one of five types of benefits available to employees in most companies. To understand employee services, an understanding of its relationship to the other five types of benefits available to employees is needed. An exploration of each follows.

Legally Required Benefits

The only type of benefit to be found in every company is legally required benefits, those that are required by federal and state laws. Worker's compensation and unemployment compensation are two of the most common legally required benefits. If a company does not offer these benefits, it risks government intervention.

Voluntary Benefits

Sometimes the company will agree to provide additional similar benefits called voluntary benefits, although these benefits are often the result of tough union bargaining. Common voluntary benefits are medical insurance, group life insurance, dental insurance, and discounts on goods and services provided by the company. Although there is no legal penalty for not offering these benefits, there is usually great pressure from within the organization to offer them. Both legally required benefits and voluntary benefits are a way for employees to save money should they ever need a good or service.

Time-Off Benefits

Another common benefit also bargained for is being paid for time not worked. Extended-time-off benefits include paid holidays, paid sick days, and paid vacations. Brief-time-off benefits include time spent eating lunch, drinking coffee, or visiting the restroom. Time-off benefits are a means for the employees to receive money in addition to what they earn on their hourly paychecks.

Earned Benefits

To receive an earned benefit, one must meet particular performance standards. For instance, to qualify for a tuition scholarship, one must usually attend school in addition to performing one's job. Other common earned benefits include bonuses, awards, and savings plans. Earned benefits are based on merit, and are not distributed equally among all employees.

Employee Services Benefits

It is the final category of benefits that most concerns the employee association — employee services benefits. Common employee services benefits include the presence of a reading room, the sponsoring of a company baseball team, the printing of a company newsletter, the upkeep of a travel bureau, and the availability of health and fitness programs. This category is considerably different from the former ones because the former benefits serve to supplement or maintain an employee's base salary while the latter attempts to improve the employee's circumstances in a way other than financially.

Many companies offer benefits in all five categories; others offer less. As benefits have become more expensive to offer, some companies have begun to offer benefits in all five categories, but ask employees to choose among them. Other employers provide employees with a choice of either more benefits, or a higher base pay. Although financial benefits are becoming fewer, employee service benefits will likely increase as research continues to document their benefit to both employer and employee.

EMPLOYEE ASSOCIATIONS COMPARED WITH OTHER ORGANIZATIONS

Even though it may offer similar services, the employee association is distinguished from other service organizations in the community because it has a unique perspective of its members. For instance, the church, the college, the hotel, and the park all provide recreation opportunities that employees would be welcome to attend, but recreation offered by the employee association is unique because it focuses on the employee both as an individual and as an employee, taking into consideration the employee's job, family, and community. Although there are similarities and considerable overlap between the employee association and other agencies, the employee association is unique in both its business operations and in its programming operations, as noted in other chapters. Because of its uniqueness, there are strategies to managing and programming within the employee association that cannot be found within the operations manuals of other agencies. These strategies are outlined in detail in the remainder of this book.

Employee associations can work in cooperation with other agencies to obtain resources and programming that neither agency could get on its own. Most service agencies in the community have two purposes - to promote individual growth and to promote community improvement - and the employee association believes in these too. By working with other service agencies, goals can be reached that no one agency could reach on its own.

The employee association is a nonprofit organization, and can be compared and contrasted with other nonprofit organizations. Both profit and nonprofit organizations convert resources into goods and services, but a nonprofit or not-for-profit organization is an organization that exists for a purpose other than primarily to make money; its purpose is to provide a service that a segment of society believes is needed but cannot be effectively provided by a for-profit agency. The nonprofit *organization* relies on volunteers, and it is therefore distinct from nonprofit *groups* such as the family, which relies on physiological motivation, and the government, which relies on coercion. In the case of the employee association, each volunteer is concerned with the general welfare of the employees and, as with all nonprofit organizations, the formal agency simply orchestrates the individual volunteer's efforts.

The employee association is a unique nonprofit organization because of its focus on employees and its mission of enhancing the general welfare and socioeconomic conditions of the employee, the employee's family, and the employee's community. Other legally recognized nonprofit organizations have a clientele and a different mission, as shown in the following lists of nonprofit organizations, their clientele, their mission, and an example of each:

1. **Community action:**
    - Clientele: citizens of the community
    - Mission: to improve the environment of the citizens of the community
    - Example: neighborhood crime watch

2. **Consumer:**
    - Clientele: all consumers and potential consumers
    - Mission: to promote wise consumption and fair trade practices
    - Example: consumer advocacy groups

3. **Disadvantaged:**
    - Clientele: the racially, economically, physically, or socially disadvantaged
    - Mission: to promote equality
    - Example: anti-poverty organizations

4. **Education:**
    - Clientele: students and those concerned about students
    - Mission: to promote education
    - Example: school boards, alumni association, adult basic education

5. **Environmental:**
    - Clientele: those wanting to save the environment
    - Mission: to maintain the natural environment
    - Example: park board, recycling

6. **Fund raising:**
    - Clientele: underprivileged

    • Mission: to raise funds to aid the underprivileged
    • Example: United Way, United Negro College Fund

7.   **Health:**
    • Clientele: those sick or potentially sick
    • Mission: to promote health
    • Example: safety groups

8.   **Information dissemination:**
    • Clientele: those needing/seeking information
    • Mission: to provide information
    • Example: library board, crisis hotline

9.   **International:**
    • Clientele: a particular nationality or all citizens in general
    • Mission: to promote a culture and/or culture in general
    • Example: cultural exchange programs; Amnesty International

10.   **Learned:**
    • Clientele: those who express high academic excellence
    • Mission: to promote scholarship
    • Example: honor society, specialist association, nonprofit research center

11.   **Leisure:**
    • Clientele: all citizens with a leisure interest
    • Mission: to promote hobbies, sports, and the arts
    • Example: "little" theater

12.   **Personal growth:**
    • Clientele: anyone seeking self improvement
    • Mission: to help one improve one's self
    • Example: Girl Scouts, Toastmasters

13.   **Political action:**
    • Clientele: potential voters
    • Mission: to promote the election process and/or a

particular candidate/issue
* Example: League of Women Voters, Republican Party

14. **Religious**:
    * Clientele: all who believe and/or "should" believe
    * Mission: to spread one's beliefs
    * Example: Southern Baptist Convention, Moral Majority

15. **Social welfare**:
    * Clientele: those having trouble adjusting to societ
    * Mission: to aid in the adjustment process
    * Example: Alcoholics Anonymous

Like the employee association, all of these organizations have a constitution, a particular clientele, a mission, and an emphasis on volunteers. Some employees may also be associated with other nonprofit organizations and this overlap is useful to the employee association, for it provides a link that can be used to foster a partnership for various activities.

THE EMPLOYEE

The employee association exists for the benefit of the employees. Before beginning to orchestrate personnel and resources within an employee association, there must be an understanding of the clientele—the employees.

Employees are constantly changing. Jim Jones' hair was black five years ago, but today it is grey. Tom White was paying off college loans five years ago, but today he has money to spare. Sally Smith was single five years ago, but today she is married and expecting her first baby. As people change, their needs change. For instance, Sally Smith will soon have a need for child-care services that she never had previously.

Psychologists note that life is a cycle, and that all people generally face similar challenges at each stage. Young employees want to maintain control over their own destinies, and yet they also want to connect with other people so they will fit in with the company. This age group is also interested in attracting a mate. Those young-married employees are beginning to raise a family. Middle adults, meanwhile, have children in college, and must adjust to living without children at home, while they are also pondering retirement plans. Although this life pattern is typical for employees in general, some do not completely match these expectations. Some never marry, some never have children, some divorce, and some encounter other unexpected joys and tragedies.

Although a familiarity with the natural life cycle is important in understanding employees, it is not the only cycle employees go through. Most also experience both an economic cycle and a leisure cycle throughout their lives. The economic cycle for an employee usually involves a person receiving training at a school or company while working for minimum wages; paying off school debts, but taking on greater debts such as a mortgage; a time of prosperity when money flows relatively freely, and a time when cash flow is much tighter. The leisure cycle involves both the *incorporation* of new activities with current activities and the *replacement* of current activities with new activities. For instance, participation in sports and opportunities to meet the opposite sex are both important for the single young male. For the engaged male, both sports and couple activities will be sought. The new father, however, will seek family activities, couple activities, and some physical sports. The father whose children have left home may drop many family activities but will maintain many couple activities, while beginning to dabble in personal interest activities.

All three cycles are interdependent and mesh with the activities an employee chooses during free time depending on the individual's economic status, place in the natural life cycle, and leisure repertoire. Note that if an activity is not in one's leisure repertoire, the individual will likely not participate in it regardless of economic situation or place in the life cycle. Leisure, likewise, complements one's work, which is part of the economic cycle. For example, if an employee enjoys activities involved in work, the individual is likely to continue doing something similar to the job during leisure time. For example, if one likes writing proposals at work then one may write novels during free time. Likewise, if an employee dislikes an aspect of the job, the employee will seek out the missing element during free time. For instance, if an employee believes he is not getting enough exercise at work, he will likely exercise during free time.

THE ROLE OF THE EMPLOYEE ASSOCIATION

The mission of all employee associations is to enhance the life of the individual employee. As noted in the previous chapter, employee associations were founded to increase employee morale, to develop friendships, to enhance communication among employees, and to offer release from work tensions. The employee association is based on the idea that employees can get more out of life. The reasons for this gap between what is and what could be are many: lack of visionary personal goals, lack of money, being unaware of opportunities, having poor health, and lack of motivation, just to name a few. The purpose of the employee association is to help employees better understand and control these problems. In fact, a typical mission statement for an employee association might read: "The purpose of the employee association is to foster social, educational, athletic, fitness, and recreational opportunities for employees, their families, and retirees."

Although all employee associations may share the same purpose, each can choose one of three major roles: auxiliary service, human development, or leisure education. Of these three, no one is necessarily superior to the others. In fact, the quality employee association fills all three roles well, fluctuating its emphasis from one to another as the situation requires.

Auxiliary Service Role

Most employee associations offer some type of an auxiliary service. Common auxiliary services include snack bars, gymnasiums, box offices, employee stores, fitness centers, and vending machines. Money from these auxiliary services is often used to fund other employee services. When an employee association emphasizes these services over other services such as programming, the employee association is said to have an emphasis on the auxiliary service role.

The employee associations that have an auxiliary service role emphasis tend to have a *power-oriented* atmosphere. These associations tend to place an emphasis on meeting the budget. Members of the association gradually earn positions of status, authority, and responsibility, and all members are expected to follow a formal hierarchy. As members and companies demand greater accountability from employee associations regarding how dues and funding are being utilized, more employee associations will find themselves falling into the auxiliary service role because they are creating stricter guidelines to better monitor themselves.

Human Development Role

The human development role focuses on programming for the individual. The employee is constantly changing as his life cycle progresses, and those who emphasize this role believe that the employee association should smooth this transition as much as possible for each employee. Counseling is emphasized to help individuals cope with career transitions — getting hired or promoted, training one's replacement, retiring, transferring — and family transitions - getting married, losing one's parents, having children, the children leaving home, the death of a spouse — and even day-to-day activities — finances, diets, arguments, and stress.

An emphasis on the human development role will give an employee association an *affiliation-oriented* atmosphere. Warm, close relationships between employees will be fostered, and all employees should be able to find encouragement and support. Because of the lack of emphasis on structure and status, all employees will feel accepted as equals among peers. This may sound better than the power-oriented atmosphere on first reading, but note that each element of every climate is not only a benefit but also a disadvantage. For instance, the lack of noticeable status may cause everyone to feel good, but without someone having the status of leader, the group is not likely to be very task-efficient.

Leisure Education Role

Other employee associations also favor programming over auxiliary services, but they focus on programming for a group of employees rather than programming for individuals. This programming may be as group-oriented as a sports team, or as individual as an art display. However, it is distinguished from individual programming because it is not tailored exclusively to one or two people. These programs can range from very complicated to very simple– a mountain-climbing expedition in Colorado to providing a new book in the company reading room. They also range from very structured to very informal– a weekly square dance club to offering a pot of coffee for employees to gather around to "shoot the breeze."

If the emphasis is placed on the leisure education role, an *achievement-oriented* atmosphere usually results. In this type of atmosphere, employees find recognition for excellent performance, such as doing well on the basketball team, or creating a beautiful pot in the ceramics club, but not for a mediocre or poor performance. Although the program is group-oriented, there is an emphasis on personal achievement, whether it is bowling one's best, keeping a commitment to stop smoking, or designing a pot. This atmosphere encourages risk and innovation and fosters the urge to try something new or daring. Examples of such risk and innovation, are employees repelling down a cliff, experimenting with a character in a creative writing class, or deciding whether to swing on a full-count in softball.

Combining the Three Roles

Each atmosphere is unique, just as each role is. Because each atmosphere presents disadvantages as well as advantages, no atmosphere exists in its purest form. Atmospheres can change, and if an employee association is unhappy with the atmosphere its members associate with it, the leaders of the association can adopt elements from other atmospheres to supplement or replace elements in their own. It is important to note that atmospheres vary daily, even though there may be a consistent atmosphere over time. For instance, if a secretary is cranky today, instead of her usual happy self, those around her will sense that the atmosphere is not what it usually is. Although atmospheres do fluctuate similar to the way the temperature fluctuates around a thermostat, the leaders of the employee association can control the "setting" of the environmental thermostat.

Diversity and flexibility are essential to an employee association's survival. The association exists as an entity separate and distinct from the corporation, and although its primary purpose for existence is meeting the needs and interests of the employees, it must develop strong auxiliary services to create a broad base of financial support to ensure its survival. The primary source of revenue for most employee associations — corporate funding through the commissions generated by the vending machines located throughout the plant facilities — is relatively unstable, because it may be redirected to other employee benefits such as cafeteria operation, at the whim of company management. Also, a drop in employment may affect all sources of income. Because of this uncertainty, auxiliary services have developed to meet a dual need of offering value and convenience to employees and also generating income to offset operating expenses incurred by the delivery of free recreational services. In times of financial crises, the income from auxiliary services may allow programs to continue what otherwise would have to be cut. The employee association is a complex business that must balance auxiliary and human services.

ORGANIZATIONAL STRUCTURE OF AN EMPLOYEE ASSOCIATION

Regardless of the philosophical role emphasized by the employee association, most associations are governed by an internally-elected board. An example of an employee association hierarchy is found in Appendix A and the construction of the hierarchy will be discussed in detail in Chapter Three.

Leadership officially operates through the formal hierarchy (Appendix A), but often there is an informal hierarchy as well. The formal hierarchy prescribes desired operating procedures while the informal hierarchy describes what actually happens. Generally, the official leader is formally appointed by the company or employee association board or is elected by all voting employees, while the informal leader is selected by those extremely close to the situation. The official leader has been placed into a role with defined duties, while the informal leader has no official role expectations. In many cases, the informal leader is a volunteer who originally had no intention of becoming a leader, but because of a specialized knowledge, skill, or talent, people — including the formal leader — often turned to that person to provide guidance for the group. Both types of leaders serve worthwhile purposes because both contribute to the success of the employee association. Competent official leaders generally recognize this, and are coordinators of talent rather than dictators. However, being a coordinator of talent is not necessarily easy, and to be this kind of leader may take even more planning than being an autocrat.

The formal hierarchy consists of only a few, if any, paid positions. Volunteers, then, are essential for the employee association's survival. Because money is not a motivator for volunteers, they must be motivated differently than paid employees. For an in-depth examination of utilizing volunteers, see *Volunteers in Employee Service Organizations* by Katie Heidrich, the second book in this series, and *The Effective Management of Volunteer Programs* by Marlene Wilson. In many respects, though, the administrator works with the volunteer in exactly the same way as with paid employees; therefore, no distinction will be made between the two in this book unless there is a significant difference.

SUMMARY

This chapter highlighted various philosophical viewpoints of the employee association–how association benefits differ from other benefits offered to employees, how employee associations compare to other profit and nonprofit groups, how associations fit into employees' life cycles, how roles develop in different atmospheres, and how organizational structures emerge – all of which are important to know when planning to operate an employee association. With a knowledge of the history of the employee association, and with an understanding of various philosophical viewpoints, an employee association can be organized.

3

Laying the Foundations of an Employee Association

Employee associations do not simply happen. Somewhere along the way, someone has taken a leadership role and brought employees together. The majority of employee associations have very humble beginnings and are initially often little more than friends getting together— employees getting together to exercise, bringing snacks to work to share on a regular basis, forming a company baseball team — but then gradually the fellowship became more structured. As this body of employees shifts its emphasis from a few employees to all employees, it becomes more formal in its operations, and eventually emerges as the employee association.

To become a legitimate employee association in the eyes of the employees, the company, and the federal government, data will need to be collected as to why the organization is needed. A mission statement and other documents must be formed. Many tough decisions must be made, such as what the documents should contain and whether to incorporate. Although the employees, the company, and the government may have doubts whether the employee association will survive, most will gladly help the association overcome such hurdles as learning legal regulations, determining an adequate accounting system, and finding facilities to utilize.

CREATING SIGNIFICANT DOCUMENTS

The first step in organizing an employee association is to assemble a board of directors to establish the cornerstones of the association. The board should be representative of all employees

— whether they are chosen by ballot, appointed by the company, or volunteered by their peers. The board's task is to draw up five documents: a mission statement, bylaws, day-to-day operational procedures, a personnel manual, and a trustee manual.

The mission statement specifies the purpose of the employee association. At this stage of development it should be well known, although no one may have written it down yet. The constitution and bylaws (Appendix B) specify procedures for electing board members, how board meetings will be organized, and how long board members' terms last. The day-to-day operational procedures outline the organizational hierarchy (Appendix A) and specific job descriptions (Appendices C and D). The personnel manual describes day-to-day behavioral expectations for the employee, such as grooming and conduct, and sometimes contains a broad overview of all of the association's jobs. Finally, the trustee manual goes beyond the bylaws and specifies practical "how to be a good trustee" rules and advice.

Each of the five documents is important. The mission statement determines the philosophical boundaries of the employee association. The bylaws determine the legal boundaries of the employee association. The day-to-day operational procedures determine the administrative pattern, and job descriptions outline administrative duties. The trustee manual, while unessential for day-to-day operations, helps maximize the potential of the board.

Mission Statement

In the eyes of the federal government, the employee association is providing a valuable service to society, and since it does not purposely make a profit for its members, it is therefore worthy of the title *nonprofit organization*. Because the federal government interprets it as a nonprofit organization, the employee association is likely eligible to receive tax breaks. However, even though it is a nonprofit organization, employees cannot make donations to it and expect to claim a tax deduction on their income tax statements. For an analysis of this subject and the tax breaks an employee association might be eligible for, see *Revenue Generation and Management* the third book in this series.

The qualifications of a nonprofit association are two-fold. The first characteristic is that it is not to exist primarily for the purpose of financially enhancing its leaders. Although the primary purpose is not to make money, most employee associations do take in a considerable amount of money from sources such as the company, vending machines, membership dues, and employee stores. The employee association budget does ideally end up with a small remaining balance, but unlike for-profit businesses, all of the money is put back into the organization.

The second characteristic of a nonprofit association is that it must provide a service to a segment of the public. Whom is this service provided to? Is it for the employees, employees and their families, employees and retired employees, or the entire community? Each employee association may answer this question differently, and their mission statements reflect these differences. Regardless of how the question is answered, it needs to be answered clearly, for its ramifications echo in everything the employee association does.

If the typical employee were asked what the mission of the employee association is, he might say to promote a better quality of life. To promote a better quality of life alone is not a good mission statement, because it does not specify for whom. Specifying for whom is important. If the federal government is to grant a tax exempt status to the employee association for providing a service to the public, it will insist upon knowing what public the association is serving. The question of "for whom" must not only be answered to please the federal government, but also to determine programming emphasis, items for which to budget, the amount of staff needed, relationships to stakeholders, representation on the board, and goals.

The following is a typical mission statement: "The purpose of the employee association is to foster social, educational, athletic, fitness, and recreational opportunities for employees, their families, and retirees." Note that it specifies broadly what it intends to do — to foster social, educational, athletic, fitness and recreational opportunities — and that it clearly specifies for whom — employees, their families, and retirees.

Constitution and Bylaws

The Constitution and bylaws are a legal agreement, and as with all legal agreements, they should be scrutinized by a lawyer. Care must be taken in drafting these documents, for problems that "will never happen here, let's not bother with writing about them" sometimes do happen. For instance, no one might think it will ever be necessary to fire a director, but sometimes it is. Likewise, no board member might seriously believe an employee would sue the employee association, but it is a possibility. Airtight bylaws can save a lot of time, attorney's fees, and effort wasted spent haggling over loose phrases in court. A sample constitution and bylaws are found in Appendix B.

Day-to-Day Operational Procedures

The day-to-day operational procedures establish a chain of command. All employee associations need a leader, a person who assumes responsibility for both the personnel and the facilities. To be successful, this leader needs to integrate workers and advisers into a successful system called a hierarchy. The number of levels within the hierarchy will vary depending upon variables such as the size of the workforce, the tasks to be accomplished, and the philosophy of the employee association towards workers.

A hierarchy consists of two types of workers: line workers and staff workers. The "line" is the people who are directly responsible for the product. For instance, at the snack shop, the line for selling candy could stretch from the cashier to the shift supervisor to the store manager to the employee association director to the employee association president. The "staff" consists of any member of the store who is not directly responsible for the product. In the snack bar setting, the staff might be the sanitation engineer whose duties are cleaning and general maintenance, and the assistant manager who handles accounts and advertising.

The importance of the line is obvious: since it is directly responsible for the production of services offered by the employee association's store, line personnel are essential if the product or service cannot be produced by the manager alone. Although staff are unnecessary for the production of the product or service, they are often necessary to alleviate burdens from the line. For instance, the line supervisor may be so busy producing the product that she does not have time to interview prospective employees, so the task of interviewing prospective employees is assumed by a staff person.

Care must be taken to avoid conflict between the line and the staff. In many cases, the line-staff battle is disguised as a volunteer-staff conflict. The two sides often blame each other for failures, with the line crying "Dumb idea" and the staff replying "You didn't try hard enough." There may be truth in these claims, because when the line and staff do not communicate well, faulty decisions are often made by the administrator. To improve communication, joint meetings might be held and a suggestion box installed. These moves lead to better decisions because of better information, and to a greater feeling of ownership in the final decision by both the line and the staff because of shared input. The line and the staff are interdependent on each other, and mutual respect should exist between these two elements of the organizational chart.

Job Descriptions

For each position, regardless of whether it is line or staff, full- or part-time, voluntary or paid, a job description must be created. A job description should include the title of the job, name of the direct supervisor, duties, necessary qualifications, and benefits. Job descriptions of a director and an intern are found in Appendices C and D.

For creating a job description, the first consideration is the responsibilities the position will entail. Next, list the skills necessary to fulfill those duties. Once the needed skills have been determined, the benefits of the position—including both tangible benefits such as money, and intangible benefits such as experience—should be listed. Next, the position should be placed in the formal hierarchy by determining the type and extent of supervision needed for the position. Once the job is placed in the hierarchy, it should be given an appropriate title.

Only after the job description has been completed should candidates be sought. Resumes should be tailored to the specific job description, so those received prior to the creation of the position will be too generic to tell if the candidate is qualified for the position. A generic resume should not influence the planning process, even if it belongs to a close friend. The only exception to this is if there is an employee with special talent in the association and a position can be formed around that talent. For instance, if the association has been using freelancers for artwork but an employee has expressed an interest in doing the artwork, a position can be created for the employee based on the association's needs and the employee's capabilities.

Personnel Manuals

The personnel manual covers a wide range of topics, including fighting, using employee association equipment, smoking, and punctuality. When establishing bylaws and personnel policies, many employee associations pattern them after the sponsoring company's personnel policies.

A policy can be defined as the standard rules of behavior that should normally be followed. Because there are exceptions, policies are usually not called rules. A policy may cover one of many controversial topics, and policies are usually concerned with day-to-day business conduct. Policy-making can be examined from both a philosophical and a practical viewpoint.

Three different philosophical approaches to policy-making exist, each with its own advantages and disadvantages; these are the incremental approach, the idealist approach, and the realist approach. Because each has advantages and disadvantages, no one approach is necessarily superior to the others, and each policy maker must decide, after understanding the pros and cons of each, which is the best option for his style of management.

Incremental approach

The incremental approach to policy making has nicknames such as "the band-aid policy" and "the squeaky wheel gets the grease approach." In the purest form of this method, no policy is established until after a need arises, and as soon as the stimulus for the policy disappears, the policy will no longer be enforced even though it may remain on the books as an official policy. Although this method of policy making has little structure and no long-range foresight, it has the advantages of being very flexible, being able to respond to a need very quickly, and being very adaptable to the wills of those with political power.

Idealist approach

At the opposite extreme of the incremental approach to policy making is the idealist approach. Those administrators who use this approach attempt to have a well thought-out policy for literally everything and have carefully examined every possible loophole in each policy.

Although this is the ideal way to plan policies, in reality it fails in several ways. To begin with, it is rarely possible to think of every problem situation and to write a policy that thoroughly covers it. If policies are planned with this depth, and the minute detail is shared with those who are affected, the policies become tedious, burdensome, and overwhelming. The creation and implementation of new policies takes a long time because of the high quality demanded of the final product. The process takes so long due to red tape that the new policy may not be decided upon until the crisis inspiring it has passed. Although the idealist policy-making philosophy is not always practical, those who strive to follow it usually have a very smoothly running operation because they resolve potential problems long before they form.

Realist approach

The realist perspective combines the best elements of the other two policy-making philosophies. In this method, all realistic policies are created, but not all conceivable policies are stated, since they are deemed unnecessary. This provides more structure and long-range planning than the incremental method, while keeping the policy-making process relatively simple. Because emergencies do arise, new policies often have to be created quickly, and the realist philosophy provides the needed flexibility. The realist philosophy also allows flexibility in enforcement, making the "logical exception" relatively easy to implement.

The realist approach is not perfect in all situations, however. Its biggest fault is that people have different opinions about what realistic policies are, and some will make too few policies, while others will create far too many. The policy-making board is constantly pulled in two directions, one side wanting more structure, and the other side demanding less. This conflict often creates disruptive power plays at policy board meetings.

Steps in Policy Making
Regardless of the philosophy of the policy maker, the process involves four steps: identifying the stimulus, creating the policy, interpreting the policy, and reviewing the policy.

Identifying the stimulus
Prior to the creation of a policy, a need for it must be identified. This stimulus can be a current problem or it can be a potential problem. A policy could be created for every event imaginable, but if an event is both unlikely and unserious in consequence, then a policy probably does not need to be created. For instance, because someone will consider lighting a cigarette in a new facility, a smoking policy should be established. Roller skating down the hallway, however, is a very slim possibility, so no written policy is necessary in most cases.

Creating the policy
Once the stimulus has been identified, a decision needs to be made about how to deal with it. In general, there are three ways to handle a stimulus: don't restrict it, limit it, or ban it. For instance, if cigarette smoking is the stimulus, smoking could be allowed anywhere in the building, it could be limited to only one wing of the facility, or it could be banned from the facility entirely.

Interpreting the policy
Just as courts are needed to interpret laws, policy boards are needed to interpret policies. Although ideally every policy is written as clearly as possible, questions will eventually arise and the policy board must address them. The interpretation of the policy should be based on the intent of the policy maker, and should flow with the employee association's goals and statement of purpose. For especially sticky situations, looking at how

other employee associations have resolved similar situations might be beneficial. The trend that has been set by one's own policy board should also be considered.

Reviewing the policy

Times change, and the policy that was needed yesterday is today obsolete. For instance, if a policy concerns the use of a gas stove in the lounge kitchenette, and the association now has a microwave oven, the policy should be removed from the books. Policies also need to be reviewed in light of how people accept them, and whether the policies are handling the stimulus in the way the policy makers hoped they would.

Policies outline consistent behavior expectations and are needed for smooth-flowing community life. Policies provide direction for handling controversial situations, they outline expected behavior of all users of the programs and facilities, and they encourage equal treatment of all patrons.

Trustee Manual

The trustee manual consists of the mission statement and the bylaws, but also contains other important information for trustees and would-be trustees. This manual may also include goals and objectives for the employee association, minutes of recent meetings, planning documents, a job description for each trustee and of trustees in general, a history of the employee association, the annual report from the previous year (Appendix H), a list of committees, and a list of current trustees with their addresses, phone numbers, and the date their terms expire. Another purpose of the manual is to provide easy access to information for the trustees, and to remind them of the mission and duties they are to accomplish.

ESTABLISHING DUTIES OF TRUSTEES

Besides creating the five documents, the board must also outline the duties of a trustee. Trustees in an employee association have five basic duties: to ensure that the bylaws are being followed, to establish fiscal policy, to provide adequate resources to accomplish the mission, to work with the chief executive, and to represent their constituents.

Protecting the Bylaws

As previously noted, the federal government has deemed that the employee association is for the public good by granting it not-for-profit status. Since the public welfare is at stake, someone must oversee the employee association to make sure that it abides by its charter and fulfills its mission. The board of directors has been given this task, and the directors are held legally responsible for ensuring that the association is abiding by its constitution. When one agrees to become a board member, one is undertaking an important trust. Being legally responsible for the employee association should not be a frightening prospect, but its importance should not be minimized either.

Establishing Fiscal Policy

Establishing fiscal policy includes outlining how cash flow will be supervised, and how the budget will be approved. Some budget questions, such as those concerning inventory, pricing, and purchasing, apply only if one is operating an employee store or snack bar, while others apply in all situations. Common fiscal questions include:

- Accounting – What type of process will be used? How often will workers be paid? How often will records be updated?

- Equipment – Can quality equipment be purchased? Is the facility able to house the equipment?

- Hours – How many hours will the leaders be available each week? Will the company store be open when the company is closed?

- Inventory – What should be sold? How much should be kept on hand? How should the inventory be monitored?

- Payment - Will the employee association's bills be paid by check? Will the bills be paid weekly? Who authorizes the checks? Will a petty cash fund be kept? Will the treasurer be bonded?

- Pricing – Should prices include a fee for labor? Should prices be reduced as a service to all employees or just to those who have paid their dues to the employee association? Should the employee association subsidize products at the box office and company store? Should a fee from each sale go to the employee association's treasury?

- Purchasing – Should items be purchased only for vending machines or will company store facilities also be available? Should name brands be purchased? Should buying be done in bulk? Should a competitive bid process be established?

- Security – Can the property be secured from thieves? Is the property safe from pests such as mice and cockroaches, and is it protected from natural disasters such as floods? If something is stolen, how is it accounted for in the financial journals?

- Services–What services should be offered? Should financial value be placed on these services?

- Renting–Should the employee association rent equipment? What should the employee association own?

Each question can be answered several ways. For instance, if products are purchased in bulk, a sizeable discount might be obtained but this may cause a surplus that will not sell for several years, tying up large reserves of cash. Different answers are correct depending on the situation, and the board will probably need to consult a wide variety of sources, including other employee associations, the director or leader of the employee association, company management, the accounting department, NESRA, and potential clientele.

Providing Fiscal Resources

Besides determining fiscal policies, the board must also approve a budget. Often this budget is drawn up by the leader of the employee association, but once the board determines the budget, it is as responsible for the contents as the administration is, so careful screening of each item is required. Periodic reviews should take place during the year so that the budget's success can be evaluated and the budget document can be modified. An example of an income and expense statement is found in Appendix E, and details of the budget document are given in Chapter Six.

Working with the Chief Executive

The board is composed of voluntary members, each of whom has other commitments besides the employee association. Because the tasks of day-to-day administration are too great for board members to perform efficiently, they select a person to lead the employee association. The extent of the board's power to hire and fire varies from association to association, and sometimes the board may only recommend a candidate.

The person selected to lead the employee association as the executive head is often a professional administrator from outside the company, and is given a full-time position. Sometimes, though, it is simply a volunteer from among the current employees. Regardless of whether it is a paid or unpaid, professional or voluntary position, board members must select the person, monitor their performance, and decide to keep or terminate the leader. The leader is the most visible of all association members and often speaks on behalf of the board and the employees, so care must be exercised when choosing this person. A leader's job title varies considerably from association to association—activities manager, human resources director, presiding officer, general superintendent of recreation, personnel liaison, and vice president are just a few, and the duties performed by the leader also vary greatly. A job description for a director is found in Appendix C. Note that the administrator's job is to carry out the policies established by the board, and not to replace the board as the legislative authority.

Representing Constituents

Because the board is representative of all employees, its members need to maintain constant contact with all employees so they know what the employees want and need, and so the employees know what is happening. Much of this pulse-taking can be done informally on the assembly line, in the office, or over coffee. It can also be done in more formal ways, such as a needs survey, or by a post-program evaluation form to find out what employees did and did not like about a recent program.

THE INCORPORATION QUESTION

The third major problem a board must grapple with is whether to incorporate. Incorporation is necessary to be a truly not-for-profit organization. This entails both federal and state recognition of a public purpose being served by a private organization. Legally, all nonprofit organizations are to be incorporated, but many are not, and the law is not enforced for most of these organizations with budgets of less than $10,000 annually.

Whether to incorporate is a major decision that must be made in the early stages of formalizing the employee association. To understand what is best in a particular case, tax and law experts and company officials should be contacted for advice. In general, incorporating has the following advantages: (1) it limits stock loss and there is no personal liability; therefore, if the employee association is sued, individual members are in no danger of losing their homes or cars; (2) several corporate tax deductible employee benefits such as health insurance will exist for personnel paid by the employee association; and (3) any surplus of the employee association can be held within the corporation without being taxed; typically this is reinvested within the association.

Although incorporating does have significant advantages, it also has major disadvantages: (1) because of lawyers' fees and other legal expenses, it is expensive to set up; (2) an annual license fee must be paid; (3) the corporation is subject to far more red-tape and regulations than noncorporations; and (4) it is much more difficult for a company to write off a corporation as a tax loss than it is to write off a noncorporation.

The decision of whether to incorporate or not is not an easy one. Several levels of incorporating exist, and each of these has some advantages and disadvantages. Employee associations generally struggle in their first couple of years, and therefore may not want to incorporate until there is a solid financial basis; in this way, the company can take advantage of tax laws while supporting the employee association. No employee association is in exactly the same set of circumstances as any of the others, so a careful inventory of the situation is important. Once an employee association has decided to incorporate, the process is as simple as drawing up the required bylaws and statement of purpose, filling out the proper forms, and paying the proper fees. A more detailed examination of the Internal Revenue Service (IRS) restrictions, tax benefits, and a step-by-step process for forming an employee association based on IRS guidelines can be found in *Revenue Generation and Management* by John Schultz, the third book in this series.

Officers

The fourth crucial decision already made in the construction of the constitution but reaffirmed at this point, is the establishment of officer positions. If the association incorporates, there must be at least three officers: a president, a treasurer, and a secretary. The bylaws of the particular employee association will determine the officers' exact titles, powers, and duties. In general, each of these are functional positions, so appointing token office holders should be avoided, although token offices can be established beyond these functional offices. Each office has prerequisites expected by the government when one incorporates — the president should command respect and should have previous experience on an employee association board, the treasurer needs to be skilled in finances, and the secretary must be able to record information accurately. The bylaws of a particular employee association may be even more stringent about the prerequisites.

SUMMARY

The founding of the employee association requires a lot of tough decisions and paper work. However, each decision and each piece of paper has a purpose and aids in giving direction to the employee association. All subsequent plans, such as the goals and objectives, refer to these documents for guidance.

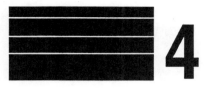

4

Organizing the Planning Process

Once the mission statement has been formed, the employee association must create a plan to give the statement specific direction. The planning process includes determining goals and objectives, assembling the planning document, implementing the plan, and evaluating the plan.

GOALS AND OBJECTIVES

The employee association needs to determine the goals and objectives that will outline how its mission is to be accomplished. A goal is a general statement of performance expectations. Goals can be either short-range, such as daily; mid-range, such as weekly; or long-range, such as five years. The first step in defining a goal is to state the general behavior the organization is to display to accomplish the task. Goals should be stated in terms of desired products and outcomes, not as processes to be undertaken. For instance, a goal might be "to obtain a membership of 500 employees."

Objectives are derived directly from goals and should specify exactly what is to be accomplished. For instance, objectives that follow the goal of obtaining a membership of 500 employees might be to recruit 200 members by telephone, 200 members by mail, 100 members by face-to-face-contact. Good objectives are composed of five characteristics, they are:

- related to the goal.
- measurable.
- outcome-oriented.

- comprehensive.
- obtainable but challenging.

Writing a good objective is an important skill in any organization, and it is a major key to a successful employee association. Well-written objectives direct the course of the association, and without this direction the association would drift aimlessly. An objective that has the five criteria discussed above serves many functions as both a teaching tool and an evaluation tool, including

1. conveying intent to others,
2. monitoring progress toward a goal, and
3. selecting evaluation procedures.

PUTTING IT DOWN ON PAPER

Goals and objectives are not the only things that need to be written down systematically on paper when one is planning — all things should be in writing for easy recovery. By writing it down you can allow yourself to consciously forget about it for the time being. The conscious mind can only hold approximately seven pieces of information at a time (Miller 1956) and writing your thoughts on paper frees the mind to think about other things.

Every successful employee association is built on a lot of unseen organizing. By putting ideas down on paper, others can verify the idea and provide feedback. Also, if one should become ill or accept employment elsewhere, others can examine the plan and study past records so they know the intention of the original plan.

COMPONENTS OF A GOOD PLAN

Planning need not be a complicated process. Good planning has five basic steps; (1) take an inventory, (2) set goals and objectives, (3) develop a specific plan that will encompass the goals and objectives, (4) put the plan into action, and (5) evaluate to see how close the goal is to being met. This chapter will focus on the first three elements and the rest of the book will examine the remaining two. Planning may be dreaded by some employees who do not have a clear understanding of the planning process or how it saves valuable time, energy, and money. Generally, though, it is welcomed by employees, because it provides a sense of direction and predicts what the future holds and why.

Taking Inventory

To begin planning for the future, the organization's past must be examined, it must be determined where the organization stands in the present, and how others perceive the future of the organization. The year-end report (Appendix H) can be used to determine where the organization presently stands in terms of members and finances. By looking at employee association records and budgets from previous years, trends and priorities from past years that are useful in predicting the coming period can be identified. Company records also predict such things as growth and down-sizing, which give insight for the future. The goals of other agencies in the community should also be considered. For example, if the community park plans to build a swimming pool, this may cause a change in the employees' free time activities.

Setting Goals

After taking an inventory, goals should be determined for the employee association. Goals are simply statements of how the organization proposes to accomplish its mission. Goals specify where the employee association should be in the future, and they should be established for short-range periods such as a month, a quarter, and a year, as well as for longer periods such as two and five years. A goal is a vision, and an ideal goal is one that causes the employee association to stretch to reach it, but it is attainable.

Once goals have been established, the association must decide what types of activities are necessary to accomplish each of the goals. To do this, objectives must be written for each goal. When the plan is fully prepared, it should be put into action. An unused plan is the same as having no plan at all.

As it is implemented, evaluate the plan to be sure it leads to the desired goals, and if it doesn't, modify it. Generally, a goal will have approximately seven objectives. If the objectives are not specific enough, one can write objectives for the objectives. When writing these objectives, keep in mind both the employee association's inventory, and employee association's goals, the first two steps of the planning process.

Developing a Specific Plan

The employee association is many things to many people. To some employees it is a store in which to shop. To other employees, it is a source of discounted concert tickets, and, to still others, it is a place to work-out. Because of this variety of uses, it may seem overwhelming to plan for each aspect but there are actually only five components to be developed: (1) the plan itself, (2) the organization, (3) the staff, (4) the day-to-day administration, and (5) the evaluation.

Planning the plan

Everything begins with the plan. Without knowing where the association stands and what resources the employee association has, without a vision of tomorrow, and without knowing how to accomplish goals, the employee association will experience stagnation or decline. When developing the plan, ample time is needed to organize, obtain information necessary to plan effectively, and brush up on effective planning techniques. Time spent in planning is an investment, and in this case the return is well worth the effort. The goal should be the light at the end of the tunnel, and even though in day-to-day operations it may seem far away, it should always be visible, providing a sense of direction.

Planning long-term administration

Planning long-term administration is merely specifying the process to be used for making decisions. This planning includes organizing voting rules for meetings, and organizing hierarchies for staff and other structures that give the employee association its shape. Although these do not need to be recreated each year, they should be reviewed periodically to see if any modification could make them more efficient.

Planning personnel roles

After planning the organization's structure, personnel can be placed in various roles. Each role should be sharply defined, and the most appropriate person should fill each position. Roles often change as technology evolves; for instance, since computers have become such an important part of everyday life, the qualifications for a secretarial position may require new skills that were unnecessary several years ago. Likewise, the abilities and attitudes of the current personnel change due to a number of factors — learning new skills at college, gaining on-the-job skills, changing ambitions, or undergoing changes that alter their lives. New staff may not need to be selected annually for each position, but personnel should be periodically reviewed — once a year is recommended—for the benefit of both the employee and the employee association, so an ideal match is obtained for both.

Planning day-to-day administration

As employees fulfill their roles and accomplish objectives, day-to-day supervision is needed to answer questions and handle unexpected crises. Planning day-to-day administration requires specifying the orchestration of resources and personnel, and specifying how supervisors will handle interruptions of the plan. Day-to-day administration should direct the employee association toward its goal, not merely provide survival tactics.

The employee association is constantly changing. This change often goes unnoticed in daily living, but becomes noticeable when recalling how many of the current personnel were present five years ago, or how much new equipment has been purchased in the past ten years. Change is ever-present, and the employee association needs to plan for it to maximize the potential of the change. It is not possible to see into the future, but educated guesses about change can help the association to plan accordingly. Long-term goals are necessary for the employee association to move forward, but it is the day-to-day operations that accomplish these goals, and if they are not planned efficiently, then it is unlikely the goals will be reached.

Evaluation
    Evaluation is the final element to be planned. It must be determined what information is needed to verify how well the plan is working, and also what the best method is for collecting that information.   The results of the evaluation should be compared to the objectives to determine if the plan's goals were met. Not all plans are good plans, and if the employee association believes it has made a mistake in implementing a particular plan, it should be modified.

SUMMARY

    Planning is a never-ending process, and it needs to be done on a   daily basis.   There are   hundreds of plans within the employee association;   for the snack bar,   the ski club, the darkroom, and so on — and all of these plans should promote the mission of the employee association. To encourage cooperation among agencies in the community and among the departments of the employee association, the various plans are often undergirded by a master plan, sometimes called a comprehensive plan because it encompasses all of the other plans. Using a master plan prevents unneeded duplication of services, saving considerable time and money while providing clear direction and vision for the employee association.
    The planning process is the same regardless of what aspect of the association the plan is being developed for. Change is something an administrator must not only learn to live with, but learn to turn to his advantage.

5

Planning for Change: Organizing Meetings and Recruitment Drives

The employee association is constantly changing. The clubs of today are not the clubs of yesterday, and the clubs of tomorrow will not be like the clubs of today. For instance, the snack facility that is so important today may be obsolete within a few years. Interests and technology are constantly changing, creating fresh opportunities while discarding others. An employee association never stops evolving during its existence.

Guidelines for an orderly transition from the present to the future are crucial. Some transitions occur periodically and can be written in a simple format, such as the guidelines for forming and disbanding clubs found in Appendix I. Other transitions are much more subtle and must be dealt with daily. This chapter examines how to capitalize on two of the biggest daily transitions —moving material through a business meeting, and recruiting new members.

CONDUCTING BUSINESS MEETINGS

As the employee association grows, the board of trustees will find that it can not govern adequately due to a lack of expertise and time. Rather than create new seats on the board, a board will often form committees composed of trustees and nontrustees who have expertise in the given area. Commonly found committees include

- finance committee: prepares budgets, reviews all financial documents, and recommends financial policies.

- personnel committee: develops personnel policies, handles violations of policies, and evaluates staff.

- program committee: reviews programs.

- nominating committee - identifies, screens, and recommends prospective trustees.

- planning committee - makes long range plans.

- special events committee - coordinates huge programs such as a charity drive or an outdoor concert that the program committee would not have time to coordinate.

- the public relations committee - it coordinates the public relations and marketing strategies.

- the building committee - it oversees facilities and the grounds around the facilities.

Employee associations do not necessarily need all of these committees, because sometimes committee functions can be grouped together. For instance, in a small employee association, the program committee might handle the functions of the program committee, the special events committee, and the public relations committee.

The business meeting is an important part of the employee association, and has been called the backbone of the association. As a member of the association, you can influence both the tasks to be accomplished, and the attitudes of the other members at these meetings.

ORGANIZING MEETINGS

Meetings are meant for making decisions, but sometimes much time is spent doing other things. For instance, the leader may talk to only one or two people, while the others present chat among themselves. A large portion of the time is sometimes spent deciding what should be discussed next. Being organized can reduce or solve these and other problems. When having a meeting, one needs to answer the following questions.

Do We Really Need a Meeting?

The first question to ask is whether a large group meeting is necessary. If the leader needs to talk to only one or two people, a private conference may be far more efficient, and sometimes even an informal meeting at the coffee machine will suffice. If the business meeting is merely a ritual and there is no business to discuss, the meeting should be cancelled.

What's On the Agenda?

If the meeting is necessary, decide on the agenda beforehand. Input for the agenda can come from each member as well as outside sources, and a centralized location should be established for adding agenda items. An outline of the proposed agenda should be circulated several days prior to the meeting, so members can brush up on the subjects to be discussed and their input can be more thoughtful and precise. Emergencies do arise, and an item not listed on the circulated agenda may appear on the final agenda at the meeting. This agenda should be distributed to each person entering the room. The final agenda should be composed of announcements and simple items at the beginning and major discussion items at the end. The agenda should be adhered to closely with no new topics added during the course of the meeting. Strictly following this latter policy may cause grumbling at first, but once members accept and follow it, meetings will run considerably smoother.

When Should Meetings Be Held?

If members have a tendency to get off-track, schedule the meeting as the last item of the day, allowing time for quality input, but not the accompanying time-wasters. All meetings should be scheduled when people are not at their individual peaks; two of the slower times for most people are immediately after lunch and immediately prior to going home, and these times are generally considered ideal for individuals to reach their collective peak. Setting a realistic time frame also helps to move the meeting along—if an agenda should be covered in half an hour, note this on the agenda, and then meet this deadline. More than once has a five-minute meeting been stretched into a hour, simply because it was tradition to have an hour-long business meeting. Start meetings on time; this tells people that the meeting should be taken seriously.

Where Should We Meet?

The meeting place also helps to set the tone. For instance, a meeting in a board room is much more likely to be formal than one in the cafeteria. The presence of certain items within the physical setting — such as ash trays, refreshments, and portable furniture — all greatly influence the tone of the meeting, since these things generally have a meaning beyond the functional one. For instance, the primary purpose of serving food is to prevent hunger pains, but there is the additional meaning that snacks may encourage friendships to develop. The purpose of each item within the setting should be examined.

What Does a Leader Do?

The leader has a great influence upon the meeting and should therefore set an atmosphere conducive to dealing with the tasks at hand. With the aid of the agenda, the leader should guide the flow of the discussion, keeping the group on the topic. Although the leader should not show partiality to a particular idea when several are discussed, the leader should be willing and able to clarify and summarize others' ideas and play the devil's advocate with all ideas. The leader should also see that the group sticks to its constitution and bylaws, and that all ideas are given a chance to be heard.

What Do Group Members Do?

Other participants also play an important role. Having seen the agenda ahead of time, they should be able to make a meaningful contribution to each topic. By monitoring their own actions, they can greatly reduce wasted time and keep the discussion focused on the matter at hand.

Should We Keep Records?

For smooth operation, the group — whether it is the board of directors or one of the many clubs — should keep records of business meetings. Record-keeping takes time, but it can save far more time than it takes. By distributing the minutes of the previous meeting to all who were present and to those who are interested but were not present, numerous verbal recaps can be prevented, and it can be assured that everyone receives the same information. These minutes help in planning future meetings, and avoiding needless repetition. Records are more than just minutes, though, and anything else that assists in the progress of the group should be saved for easy access. Often it is little things, such as the amount of money spent on soft drinks at the Halloween party, that those planning a future gathering will want to know.

Section Summary

Meetings are sometimes essential because the input of many people is needed to make a complicated decision, and sometimes many heads are better than one. Meetings, as with any administrative tool, should be utilized whenever applicable but should not be abused by overuse.

RECRUITING NEW MEMBERS

Once the goals and objectives of an employee association have been specified, thereby creating a satisfactory program, an effective communications network, and a solid organizational structure, the employee association is ready for its first membership drive.

Establishing the Goal of the Drive

The first step in a membership drive is to determine a membership goal — the number of people that can realistically be expected to join. Ideally, there will be a never-ending recruitment program which is highlighted by an annual membership drive.

Forming Committees for the Drive

A membership drive is not a one-person job, and the use of the committee process can increase the chances of success. Typical recruitment drive committees and their purposes are listed below.

- General campaign committee: serves as the clearinghouse for all matters pertaining to the campaign.

- Publicity committee: prepares the publicity program

- Program committee: develops theme and slogans and arranges for meetings

- Arrangements committee: prepares all meal functions involved in the campaign

- Prize committee: selects and buys prizes for awards to recruiters

- Auditing committee: takes care of paperwork involved in receiving applications and dues

Obtaining a List of Prospects

Once the committees are formed, a prospect list needs to be obtained and prioritized. Generally the company will gladly provide a list of employees for this purpose. Previous members, new hires, and employees with children are generally good prospects. Employees who commute far distances, car poolers, and those who rely on public transportation are less likely to join the employee association.

Determining a Theme

A membership drive theme needs to be selected to provide a focus for the drive. Themes are limited only by the imagination and can arise from sources as varied as popular songs, employee service literature, and the company cafeteria. Themes can be as poetic as "Putting the You in UMYF" to as common as "I Want You — For the Employee Association." Some employee associations will benefit from creative themes while others will succeed best with traditional themes. A theme needs to be selected with care, for both the meaning it conveys, and the ability to program the recruitment drive around it.

Soliciting Prospective Members

Once the theme is in place, the employee association is ready to begin contacting prospective members. The method of contact may be personal interviews, direct mailings, or telephone solicitations. The company may be able to make recruitment easier by subsidizing association dues, or including a payroll deduction for association membership.

Person-to-person contacts

The person-to-person approach works best for groups of a manageable size, for it requires numerous people with high levels of energy. The volunteers recruited to solicit should interview twelve to fifteen prospects and ideally these should be people they know. The volunteers should be familiar with the employee association's programs, and be able to communicate the benefits of being a member of the association. Proper training is essential for all solicitors.

This method of recruitment is useful for several reasons. Because of the personal contact and the exertion of peer pressure, the percentage of recruits is high. This method also provides a chance for new members to provide feedback about the employee association. At the same time, volunteer solicitors tend to greatly strengthen their own membership ties.

Telephone solicitations

Telephone solicitations are more time-effective than person-to-person contacts because the volunteer can make many more telephone calls within the same amount of time. This method is also far more personal than a form letter, and any question a prospective member has can be answered on the spot. Training is also important with this method. Telephone companies interested in selling phone service may be willing to present a training seminar. When properly trained, a well-informed volunteer with common sense and a pleasant voice can be an extremely effective recruiter.

Direct-mail contacts

For large employee associations, direct mail may be the best alternative. Each mailing should contain:

1. a personal letter from the president of the association
2. a membership brochure detailing benefits
3. a current issue of the house newsletter
4. a current promotional flyer or program update
5. a simple, attractive application form
6. a return envelope

Offering a membership incentive works well with all three methods, especially the direct-mail method. Incentives might include a free membership button for every new member, or a free cap to the first one hundred members who reapply. Buttons, caps, and other visible items help to promote the membership drive among other employees who see these items. Another incentive could be a drawing for a prize, with each member's name being placed into the lottery.

A second and possibly third mailing may be necessary. These mailings should follow within a few weeks of the first mailing. Typical response rates are sixty percent on the first mailing, ten percent on the second, and ten percent on the third, according to NESRA statistics. If possible, one of these mailings could be inside employee's paychecks.

Section Summary

Recruiting is a year-round job. Every new employee is a potential member. By viewing recruitment and retainment as an ongoing process, and planning for the recruitment drive months in advance, the employee association can help ensure a steady flow of new members.

CHAPTER SUMMARY

The successful administrator admits that change is a fact of life. Smooth business meetings and an on-going recruitment drive are testimonies that the employee association is serious about keeping abreast of change and that it is committed to excellence.

Tools and Techniques of Administration

The administrator of an employee association is many things to many people, so he must be familiar with a variety of duties, tools, and management techniques. The following sections are overviews of task-related items that all leaders of employee associations must deal with: (1) budgeting, (2) administrating employee services and recreation, (3) internships, (4) workshops, (5) marketing/public relations, and (6) communication. The information given here is merely an overview, but sources with more detailed information are referred to in the annotated bibliography.

THE BUDGET DOCUMENT

Budget items are not selected at random from many possibilities; they are items selected from a range of options with a particular motive, as specified in the goal, in mind. This motive, as well as an estimate of expenditures and of receipts, supplemental information, and a summary should be a part of every budget.

The Budget Motive Statement

The motive statement is often technically called the budget message. The budget is a statement of how the employee association plans to utilize its resources to meet its goals, while staying within the guidelines established by the bylaws of the employee association, the regulations of the company, and the laws of the city, state, and federal governments.

Budgeting is an important part of the planning process. To create a budget, one needs to determine the goals of the employee association, the specific projects to be carried out to meet those goals, the supplies and personnel needed to complete the projects, and an estimate of the financial costs and benefits of these expenditures. A clear motive statement helps to pave the way for the rest of the budget, for it justifies and explains all that is to come.

The motive statement generally consists of five aspects: (1) highlights of the past year, which help to create a positive atmosphere and establish trends; (2) highlights of some problems facing the employee association, and how the new budget will help solve these problems; (this shows that the employee association knows it is not ideal but that it is taking serious steps to improve itself); (3) possible problems of the future and information on how the budget prepares to deal with them; (4) programs and services that are being added, dropped, or significantly improved; and (5) information on where significant changes in expenses and receipts will be made, clearly specifying why the change is taking place. Each of these five points emphasizes that care must be taken not to rely exclusively on the present year's receipts in forecasting the next year's budget. The forecast must also consider such variables as inflation, social and economic factors, public opinion, fads, trends, and even weather conditions. Administrators will find that computers can help in the budget process, pinpointing trends that might otherwise go unobserved, and also aid in developing spreadsheets to complete projections. The board will likely ponder all five of these points, and it is to both the board's and the administration's advantage to have these items well thought-out ahead of time. Incomplete budget preparation may cause board members to suspect incompetence, which can greatly strain a board-administration relationship.

The motive statement, in many respects, is a public relations tool. It must justify the figures in the budget, and it must leave no doubt that contributions from funders — both the source that formally approves the budget, and those who contribute money toward the budget — benefit both the contributor and the employee association. Most companies and other donors want to be sure they are getting maximum mileage for their dollars, therefore the motive statement should show that the employee association offers high quality programs, provides a wide range of worthwhile services, and is very well managed. Day-to-day public relations activities should already have promoted the employee association so the funder has a positive view of the association, and would have no regrets about being associated with it. If the employee association does not have a good public relations strategy, and if the motive statement is the first major positive contact the funder has heard, the motive section may not be very influential. However, if past public relations have been successful, the summary provided in the motive section will have great impact on the funder.

Other Sections of the Budget

The budget message is followed by the budget summary or the expenditure/receipt section. These two sections are often interchanged. The last section of the budget consists of supplemental material, which might include everything from proposed publicity brochures to previous balance sheets. Each section is examined below.

Budget summary and the expenditure/receipts section

The budget message section is followed by either the summary or the expenditure/receipt section. An advantage of placing the budget summary next is that it can serve as a transition from the motive to the indepth figures. The budget summary is simply the budget reduced to outline form, comparing major expenses and receipts of the past years with the projected expenses. It generally includes the sources of funding, such as the company, dues, and the box office; sources of major expense, such as the program committee; major items being purchased, such as a new photocopier, athletic equipment, and personnel time; and a general list of funds, such as the general fund, salaries, and leases. The receipts and expenditures section breaks each of these categories into subcategories, so that specific dollar amounts can be studied. For instance, the budget summary section would list "athletic equipment" while this section would specify "baseball bats," "footballs," and so forth.

Supplemental material

The last section, which consists of supplemental material, is generally the longest. The motive statement and the summary are only a few pages each, and the list of expenditures and receipts varies in size, depending on how detailed the board expects it to be presented as conveyed in the budget guidelines. The supplemental material should provide information justifying the logic and numbers presented elsewhere in the budget. The program, piece of equipment, or salary for which the money is to be spent must also be justified. Other items often placed in this section include goals and objectives, plans, evaluation material, statistics on employee wants, and capital improvement information. This section is an appendix to the budget and should contain anything referred to elsewhere in the budget.

Section Summary

Budgets vary greatly in format, detail, and length. Some association budgets are little more than blank checks with a ceiling specified but leaving no specified amount for particular expenditures. Some contain minute detail, specifying each item; and some justify every expenditure and receipt while others simply build on what was done the previous year. Although they do vary considerably, all budgets are a statement of the organization, and an outsider can determine what the employee association's priorities are by looking at it.

Besides the budget, financial documents used by the employee association should include the balance sheet, a document that reflects the employee association's current financial standing; the income statement, which summarizes cash intake; the cash receipts journal, which records every transaction in which the association receives money; the cash disbursements journal, which records every transaction in which the association distributes money; the general journal, which records all noncash transactions; and the general ledger, in which an account by account summary is kept. Financial documents are important in making decisions, justifying programs, and proving accountability. The board of directors is ultimately responsible for these as part of its fiscal management duties, but all branches of the employee association have a stake in the financial statements.

EMPLOYEE SERVICES AND RECREATION

Although the growth of employee services and recreation programs has leveled off slightly over the past three years, the fact that they increased one hundred-fold between 1979 and 1982 Klarreich (1987) indicates that they are one of the fastest growing aspects of the employee association.

Employee services and recreation is a term used to describe a series of activities that encourage an employee to adopt a healthier lifestyle. Synonyms include *health promotion program, health enhancement program, employee wellness program, disease prevention program, company recreation, corporate fitness, corporate recreation, employee recreation* , and *industrial recreation*. Employee services and recreation are based on the assumptions that it is better to prevent an illness than to pay to cure it, and that better health can be encouraged through modifying employees' lifestyles.

Employee Services and Recreation Today

As noted in Chapter Two, there are numerous motives for offering employee services and recreation, and also for employees to attend, but employee services and recreation programs have one primary goal — to increase the employee's health while simultaneously lowering the cost of health care for both the individual and the company. Although *employee services and recreation* sounds oriented toward physical health, an equal focus is also given to mental, emotional, social, and spiritual health needs in a well-rounded employee services and recreation program.

Employee services and recreation programs tend to be very personal, and both group programs and one-to-one programs are offered by most companies. Topics covered often include personal relations, sex, stress, drinking alcohol, smoking, nutrition, exercise, and leisure counseling. Objectives of offering employee services and recreation are to replace bad habits with good habits, to improve employee health, and to reduce health care costs. This replacement is typically done by providing the employee with accurate information about the subject matter, which in turn creates self-motivation to change for the better. To aid in this change, positive peer pressure and a constructive environment for the new habit are fostered. Most employee services and recreation programs have found that self-respect leads to better health, and vice-versa.

Employee services and recreation programs are usually led by a combination of paid staff and volunteers. The company may provide onsite facilities or nearby facilities may be rented. Programs tend to be educational, physical, or social. How an individual employee classifies a program depends greatly upon the individual's needs. For instance, a square dance program may present a physical opportunity for people who need exercise, an educational opportunity for people who do not know how to square dance, and a social opportunity for those looking for new friends.

Facilities for employee services and recreation in the United States vary greatly. For instance, the Physis facilities in San Fransisco are underground; the General Electric facility in Cincinnati includes a swimming pool, indoor and outdoor tracks, stationary bicycles, weightlifting equipment, and an aerobics area; the Nationwide Insurance Company in Columbus, Ohio, has no facilities of its own, but utilizes the local YMCA facilities; and Taco Bell in Provine, California, has facilities connected to its offices. Although facilities differ considerably in size, they are generally not nearly as fancy as commercial facilities because they do not have to constantly lure clientele as the commercial facilities must do.

Employee services and recreation programs in the United States appear to be meeting some of their objectives. For instance, participants in a General Electric fitness program were absent from their jobs 45 percent fewer days in 1987 than employees who did not participate and they also expressed more satisfaction with their jobs (Shinew, 1988).

Administrating Employee Services and Recreation

Although employee services and recreation consists of many varied programs, each employee association leader has similar administrative duties regardless of the type of programs the association offers. It's easier to understand the principles behind these duties by looking at psychologist Abraham Maslow's hierarchy of needs. Maslow claimed that only when the basic physical needs of food and water are met can people concentrate on safety needs. The goal of the employee association is to help the employee become self-actualized, and that means having employees who feel that their basic needs are met, that are secure, accepted, and recognized by their peers for their abilities. Leaders of employee associations have little control over whether a person is liked and valued, but they can foster environments conducive to the employee being liked and valued. Although leaders may have only minimal control over the upper end of Maslow's hierarchy, they do have control over both the physical and safety needs of the employees. For a more indepth examination of Maslow's model applied to employees and volunteers, see *The Effective Management of Volunteer Programss* by Marlene Wilson, listed in the appendix.

To be successful at overseeing an employee services and recreation program, administrators need to familiarize themselves with five things: (1) planning, (2) activity alternatives, (3) the law, (4) safety, and (5) testing. Although each of these is examined separately below, all five are tightly interrelated.

Planning the Overall Program

Planning involves collecting information about the employees' wants and needs. It also includes forming objectives and seeing that the plan is workable. Planning should touch upon all areas of Maslow's hierarchy — food and drink should be available if employees become hungry or thirsty during the activity, the environment should be secured of hazards, conditions should be favorable for employees to be friendly with each other, the environment should foster self-esteem, and the potential for self-actualization should be present.

Creating Alternative Activities

By offering a wide range of activity alternatives, one is more likely to draw more employees who would not otherwise participate. For instance, running is very good exercise, but it is certainly not the only exercise to keep the heart healthy. Many people who cannot run can go for extended walks that produce similar results; therefore, one may want to offer both a running and a walking program. The employee association wants to reach *all* of the employees and therefore a wide variety of programs at all ability levels are generally offered.

Knowing Legal Standards

Administrators must be familiar with the law related to their activities. Legal standards exist for many types of equipment and for how the instructor-administrator, instructor-employee, and employee-employee relationships are to exist. An infraction of these laws can lead to a lawsuit of negligence, malpractice, or sexual harassment. Most employee associations have a moderate insurance policy to cover a lawsuit. However, the majority of leaders in employee associations, while acknowledging the potentiality of lawsuits, have no dread of them, because they know the law and strive to adhere to it.

Until recently, liability waivers were not enough to clear the employee association in the case of a lawsuit; primarily because judges did not think individuals realized what they were signing in a liability waiver. However, this trend is changing (Executive Update, January 1989). Having participants sign a liability waiver can do no harm, and the signature could play a major role in the defense of a lawsuit, or it could even discourage a lawsuit. A sample liability waiver is found in Appendix G.

Promoting Safety

Safety concerns are dealt with in a variety of ways. If a fire, tornado, or life-threatening injury does occur, predetermined emergency procedures should be followed. These procedures should be memorized by all program supervisors, and they should also be posted on a wall or bulletin board for easy reference.

Another important component for safety is qualified instructors and supervisors. These must be trained with both theoretical book-learning and hands-on experience. Instructors and supervisors must be thoroughly familiar with all rules and procedures. Even those with a background in their specific area of supervision will need some training because each employee association stores materials in different places, and has unique traits and procedures. Training is sometimes put off because leaders think "that kind of accident will never happen here," but taking this gamble can be extremely costly in terms of time, lives, and money.

To further ensure safety, most employee associations form safety committees. The safety committee usually has the duties of inspecting facilities, equipment, and techniques, of educating others about safety, and of investigating accidents. The employee association may also invest in insurance, having policies for general liability, employee theft, and property damage. A detailed examination of the financial aspects of the employee association and insurance can be found in *Revenue Generation and Management* by John Schultz, the third book in this series.

Testing Employees' Health

Regular periodic testing is needed for all employees to determine if they are fit to stay in the fitness program, and also if they are meeting their personal goals. Much testing can be done by self-examination, but since individual employees generally do not have the equipment or the knowledge to make a detailed indepth inspection, formal testing should also be done on a regular basis. For instance, an employee can monitor his own pulse, which helps to note whether he is relaxed, but it does not detect heart disease as an annual physical given by a certified physician would.

The Future of Employee Services and Recreation

Although employee services and recreation programs have boomed in the past decade, and are certainly more than a fad, it remains to be seen if the trend continues upward, levels out, or declines. As employee services and recreation programs have become more popular, long-term studies have been undertaken to determine if these actually benefit the company and the employee, or if this money could be better spent elsewhere. Current research (see Shepherd, 1986) suggests that employee services and recreation programs do significantly benefit the white collar worker, but the findings about the effect on the average assembly line worker are too varied to be conclusive. Not only is a lot of money being invested in employee services and recreation, but a lot of time is being spent also. Currently, participation in employee services and recreation programs is voluntary, and in most companies is done in the employees' free time. Every minute an employee spends participating in employee services and recreation is a minute that could have been spent elsewhere — at home with one's children, in church, volunteering, or watching television — and the effect of this time trade-off has not yet been examined in detail by researchers.

Other questions also haunt employee services and recreation, and will have to be dealt with eventually. For instance, the average age of the typical worker is increasing, and only time will tell if the changing work force will have the desire to participate in employee services and recreation as they are offered today. If it is true that employee services and recreation programs save the company money, and net the company far more on its investment dollar than it could make in any other way, who should receive the financial savings from this investment — employees, the company, or consumers? Another issue to consider is that as employee services and recreation become more stabilized, the activities may become more of a ritual for many, and although the employees organize their daily schedules around it, it may come to be little more than a way to occupy time. Employee services and recreation have already developed some rituals that hinder people from feeling socially accepted, so they no longer participate in it. For example, if a particular brand of jogging shoes is expected to be worn on the running track, and there are employees who cannot afford the financial investment of that brand of shoe, they may not participate in the running program. Creative planning and incentive programs have been suggested as ways of alleviating the latter problem, but the long-term effects of these are not known.

Section Summary

The field of employee services and recreation faces numerous challenges as the workforce continues to change.

However, keep in mind the programming steps arising from Maslow's theory about employee needs and remember that employee services and recreation are based on the assumption that it is better to prevent an illness than to pay to cure it, and that better health can be encouraged through modifying employees' lifestyles. With these in mind, the employee association will be able to adjust employee services and recreation to fit the needs of the employees regardless of how the future changes the workforce.

INTERNSHIPS

Internships are an investment in the present and the future for both the employee association and the intern. An internship can benefit both, not only financially, but also educationally.

Structuring the Internship

An employee association needs general guidelines for supervising and rating interns, and in knowing what tasks an intern will be expected to accomplish to receive credit for the internship. The exact requirements of the internship should not be specified until the intern and the supervisor of the employee association have negotiated them face-to-face. Each intern desires slightly different outcomes from the experience and because all interns possess a different range of skills, it is important to fit the intern with the appropriate position.

The Association's Obligations

An intern should not be treated as a source of cheap labor and assigned trivial work or janitorial tasks. Most interns have several years of college experience, and have developed skills that will go to waste and deteriorate if the intern is assigned unchallenging work. Although it is important for an intern to experience lower-level jobs to better relate to people in these roles, the talents of the intern should be tapped and challenged. An intern probably possesses a skill that no one else in the employee association has, and this potential should be uncovered.

The Intern's Obligations

A prospective intern should look at the employee association carefully to see if a position there truly provides an opportunity to gain the experience and skill-level desired. The prospective intern should seek a meaningful internship because secretarial or janitorial work mean little on a resume, even if completed at an agency the intern is interested in eventually working for. The ideal internship should also present the intern with the possibility of failing — not to the degree that it ends a career, but to the extent that the intern is able to learn and grow from mistakes. Once the intern agrees to work for an employee association, he is expected to behave as if he were a full-time employee — showing up to work on time, meeting deadlines, and following conduct codes.

Results of an Internship

An internship should be a learning experience for both the intern and the employee association. At the end of the internship, the intern should have an increased knowledge of employee associations, and how to operate them. The employee association should have benefited from the specialized talents of the intern. Both should possess a copy of a written project by the intern. This project comes from the intern's interests as well as from the employee association's needs. For instance, if the intern was interested in organizing clubs and the employee association had no official guidelines, the intern could assemble a how-to manual on organizing a club. This project would not only provide the employee association with a tangible resource, it would also provide the intern a quality product to present when interviewing for a future job.

Section Summary

The ideal internship is a win-win situation for both the employee association and the intern. If the internship benefits only one of the parties or neither — as is sometimes the case — the internship process needs to be reviewed. Interns are the future leaders of the employee association, and are a resource worth developing.

WORKSHOPS

The workshop can be approached in many different ways. To some administrators, the workshop is simply another business meeting. For others, it is a part of the corporate wellness program. And to still others, it is visiting with consultants to solve problems. Producing an effective workshop requires examining the purpose of workshops, the selection of participants, the coordination of the workshop, and how to build teamwork among diverse viewpoints. Regardless of how a workshop is classified, what is most important is the end result.

The Purpose of Workshops

The purpose of a workshop is to work on pertinent problems with peers who share these problems, and with consultants who can offer additional insight. The problem can be either personal, such as a group conflict, alcohol abuse, or leadership training, or it can be a collective problem, such as the mounting employee association debt, designing a new facility, or increasing membership in the employee association. The peers present are those who share the same problem, and although there may be a wide range of rank within the employee association, all present at the workshop are treated as equals. The consultants present may be national experts, or they may be specialized citizens from the community, and they too are treated as equals. Although the workshop is opened on a note of formal introduction, it should quickly generate into a discussion with the consultant serving as a facilitator and resource person.

Selecting Participants

Choosing who should be present at the workshop can be a touchy matter. Certainly some of the employee association's leaders should be there, because their presence is supportive of the workshop, and they probably have some practical advice since they are so close to the situation. When choosing consultants to invite, their backgrounds should be examined to verify that they are knowledgeable on the workshop topic. Because of limited seating or a limited materials budget, potential employee participants may have to be screened. Criteria used for screening employees may include how much benefit the employee can obtain from the workshop, how much the employee can contribute to the workshop, how prepared the employee is to deal with the topic, and how enthusiastic the employee is about attending.

Coordinating the Workshop

Coordinating a workshop is generally handled by a planning committee. Besides selecting the consultants and determining admissions, the planning committee also has the duties of finalizing the topic, stimulating interest among employees, forming a budget, securing facilities to house the workshop, organizing a staff and other committees, obtaining materials, handling problems that arise during the workshop, and providing any necessary follow-up. Rather than do everything itself, the planning committee often appoints other committees, such as the social committee, which plans for games and refreshments; the evaluation committee, which reviews the meeting and solicits input for improving the next workshop; the hospitality committee, which welcomes guests and takes care of their needs; a decorations committee, which provides an atmosphere by making the room attractive before the workshop; and the clean-up committee, which returns the meeting room back to its original form.

The planning committee also has responsibility for determining a schedule for the workshop, but in smaller workshops only broad guidelines are presented. To outsiders, this may appear to be shirking duty, but actually it fosters the workshop spirit of teamwork. By having participants create their own schedule, one that best meets their needs is adopted. Keeping the schedule flexible also allows the participants to do what they believe is best for them.

Developing Teamwork

This spirit of cooperation is one of the benefits of the workshop. In addition to examining problems in detail, workshop participants also learn to work as part of a group. A spirit of democracy pervades the workshop, because all ideas are subject to comments from everyone — the consultant, the leader of the employee association, the assembly line worker, or the office manager. In fact, the wider the variety of viewpoints, the better an understanding of the problem the participants can obtain.

Section Summary

The workshop is not the answer to all problems. It requires a time investment from all participants, a lot of planning, and sometimes a large financial investment for overhead and consultants' fees. When used properly, however, the workshop is an excellent tool for both problem solving and building teamwork.

MARKETING/PUBLIC RELATIONS

Marketing and public relations are not one and the same, even though they have much in common. The purpose of marketing is to get employees to participate in the association's programs and to use the association's facilities. Public relations' purpose is to give all people with a stake in the employee association — the employees, the company, the community, and other employee associations, to name a few — positive but realistic feelings about the association.

Philosophy of Marketing and Public Relations

Both marketing and public relations can be approached from one of two ways: from the top down or from the bottom up. In the top-down marketing philosophy, the leaders of the employee association plan a program and then try to convince employees to participate in it. In top-down public relations, the employee association goes about its daily operations and the public relations representative explains these happenings to the various stakeholders. The bottom-up philosophy is quite different. In a bottom-up marketing approach, the leaders of the employee association investigate what the employee association members want and then build programs to meet these wants. In a bottom-up public relations approach, the leaders first pinpoint the various stakeholders' wants, encourage the employee association to move in that direction, and then attempt to explain how the employee association is indeed moving in the direction the stakeholder desires. The bottom-up viewpoint has gained popularity in recent years, and because it emphasizes people over product, employee associations will benefit best from it.

Methods of Marketing and Public Relations

Tools used in marketing and public relations functions are limited only by one's imagination: the office intercom, posters, slides, photographs, testimonies, a cafeteria booth, the company newsletter, local radio, television, and local newspapers are just few of the many mediums available. Although these options may appear to be fairly standard, announcements can be altered to receive special attention. For instance, the bulletin board may be so cluttered with posters, that few people would read your announcement there, but a poster in an unusual place might receive special attention. Likewise, announcements over the intercom may become ritualistic and virtually ignored, but if an announcement is preceded by twenty seconds of Christmas music in July, people may perk up and listen. Marketing and public relations should not always be flashy because soon the flashiness may become expectedand people begin to take it for granted. Marketing and public relations must be unique enough to grab and hold attention without being in bad taste.

Having a good message is not enough for successful marketing and public relations. For instance, if the leaders assert the viewpoint that fitness is important, and yet they themselves are inactive, others are not going to accept their message about fitness. In a sense, each leader is a walking advertisement. Not only should leaders represent what they say, they should also know the facts. A poorly researched viewpoint will usually be detected by others, causing a loss of credibility.

Formal sources such as news releases, newsletters, and broadcast copy are an important part of passing on both public relations and marketing. However, informal sources such as coupons, tee-shirts, buttons, and ballcaps are also important. The informal source of word-of-mouth is often the most efficient and cheapest way of promoting the employee association.

Section Summary

Marketing and public relations cannot make a bad program good. An excellent public relations/marketing campaign can get results temporarily, but people quickly see through the campaign to the program itself. Public relations and marketing can supplement a good program by both attracting and retaining people, and they can boost the image of the employee association.

COMMUNICATION

Although some tasks require working with a large group of stakeholders, others require working in a small group setting, and still others require working alone. Communication, both verbal and nonverbal, is important in all situations, especially in one-to-one settings. In larger groupings there is always someone the message receiver can turn to and ask, "What did he mean?" but not in a one-to-one setting. Communication should be analyzed by its form - verbal or nonverbal - and by its process - decoding and transmitting.

Verbal and Written Communication

Language is an important tool in communicating. The largest part of language is grammar, but it is far more than words alone. Language is the tone of voice one uses to convey the meaning - for instance, "Sam drove to the *store*?" "Sam *drove* to the store?" and "*Sam* drove to the store?" all imply different questions - and the style of handwriting with which the words are written — for instance, in the previous example special attention is given to the italicized words.

Non-verbal Communication

Not all communication is verbal or written language. Other methods of communication include

- eye contact: insecure people often look at the floor.
- sitting position: couples dating who want to touch sit

side by side, a leader-subordinate pair will sit at right angles, and two equals will sit across from each other.

- touch: leaders touch subordinates more than subordinates touch leaders.
- smiling: people smile when they are friendly regardless of whether they are happy
- gazing: people gaze at those whom they like much more than those whom they are neutral toward or whom they dislike.
- body posture: a tense body indicates that one is not relaxed.
- distance: intimate friends stand close together; casual friends stand within five feet of each other, and general acquaintances stand further than five feet apart.

Each of these presents a message that complements the other messages one is sending.

Transmitting Messages

Employee associations generally use three methods of sending messages: in person, by telephone, or by memorandum. Personal messages tend to get faster results, but they also may require interrupting the work of others. Written messages tend to be superior in terms of time invested because a message takes two minutes or so to write and seconds to send. Personal contact often appears to take less time, but in reality it involves walking to the office/assembly line, waiting for the other person to finish a task, making small-talk, presenting the message, making more small-talk, and walking back to the work area. It is also more time effective to write one memorandum to twelve people than to try to talk to each of them personally. Telephoning has some of the advantages and disadvantages of both — it does interrupt the other person, but it does save walking time, and the time spent making small-talk is usually less. The best method depends upon the immediate circumstances. Delivering the message personally may be more important than the time you save doing it by phone.

Decoding Messages

Communication can break down in the decoding of the message as well as in the transmission of it. Because a message must be decoded, the sender should compose as simple a message as possible. In other words, memorandums should include all pertinent information in an easy-to-decipher format. To this end, many employee associations utilize a memorandum that requires appropriate checks, and leaves room for a sentence or two of additional information. In speaking and writing, all ideas should flow in a logical progression, and all jargon that could be misinterpreted should be avoided. Keep in mind the viewpoint of the receiver, and try to build a two-way bridge between the receiver and the sender of the message.

Section Summary

The communication process can help or hinder almost every situation in the employee association depending on how it is utilized. Whenever people come together they communicate facts, ideas, and feelings, and sometimes the things they don't say send the most memorable messages. Both the sender and the receiver must clarify and confirm what is being communicated if the message is in any way unclear.

CHAPTER SUMMARY

Budgets, employee services and recreation, internships, and workshops are all aspects of an employee association over which an administrator has much control. Each of these tools can further the employee association toward its goal. Each also helps in promoting favorable relationships with other organizations by clearly specifying what the employee association values and what it is doing to overcome any problems it has. This relationship with other organizations is explored in depth in the next chapter.

Working With Stakeholders

The employee association is one organization out of hundreds within a city, and since no organization can completely isolate itself from the others, the employee association must work with a variety of other groups. These groups may be as large as the sponsoring company, or as small as a local one-person vending company owner. Each of these groups has a vital interest in the employee association, and may even possess the power to make or break the association. Because each of these individuals and organizations has a stake in the future direction of the employee association, they are called stakeholders.

Most stakeholders see the employee association from only one perspective — the way the employee association affects them. They may make assumptions about the employee association that are rarely questioned even though they may be incorrect. Even though the stakeholders may hold misconceptions about the employee association, they are an important power and cannot be ignored by the association's administrative team. To deal successfully with stakeholders, it helps to understand how stakeholders make their assumptions, to identify the stakeholders and their relationships to the employee association, to classify stakeholders, to solicit input from the key stakeholders, to combine all stakeholders' interests into a strategy, and to share the strategy decisions and logic behind them with the stakeholders.

PINPOINTING ASSUMPTIONS

Assumption-making is human nature. With only partial information, people seek to complete the picture with their own ideas. By making assumptions, people are able to categorize the world around them, changing their world from abstract to concrete. Assumption-making begins at an early age and continues throughout life. Often it is only when an individual's assumptions contradict someone else's that the assumptions are even consciously recognized. For instance, when listening to the announcements on the intercom, a person may subconsciously put a face to the voice of the announcer and not question this assumption until he actually sees the person.

Employee associations are more concrete than a voice on the intercom, but stakeholders still imagine the employee association to possess many traits it does not have. The question is not whether stakeholders have assumptions, since they obviously do; the question is whether these assumptions are correct, and if not, how dangerous to the employee association is it if the assumption is not corrected?

IDENTIFYING STAKEHOLDERS AND
THEIR RELATIONSHIPS

Stakeholding is often a two-way street, with the stakeholder and the employee association receiving benefits from each other. Although all employee associations have similar categories of stakeholders, the strength of the relationship with a particular stakeholding group varies depending upon circumstances. The following is a list of common internal stakeholders and a brief description of the employee association's stake within them.

- Accounting - provides tax information, standardized accounting forms, and answers financial questions

- Communications - assists in writing/printing invitations and newsletter articles

- Company administrators - work as peer advisers to the employee association administrators, and often loan facilities to the employee association

- Data processing - prints labels, keeps track of fees, handles billing, sends invitations, and categorizes employee association members

- Employees - pay dues, volunteer their time, and participate in programs

- Human resources - provides data about prospective members of the employee association and often aids in answering questions about employee behavior

- Legal - answers questions regarding policy making and liability

- Marketing - helps promote special events and the "good name" of the employee association

- Medical - responds to emergencies and is at special events such as the annual football game

- Physical plant - helps set up and clean up facilities, and solves plumbing and heating problems

- Public relations - helps to promote a positive image to all other stakeholders

- Research - assembles survey instruments and collects and analyzes data

- Security - patrols the building, supervises parties, and responds to emergencies

External stakeholding groups often include the following.

- City council - passes building laws and codes regulating the company and the employee association

- Commercial establishments - encourage the formation of new facilities which will complement their own

- Environmental groups - promote the formation of parks

- Other employee associations - provide friendly athletic competition at company ballgames and experiment with new ideas

- Schools - supervise college interns and promote after-school programs for elementary students

- Venders - restock the snack bar, company store, and vending machine

- Volunteers - provide labor and helpful ideas at little or no cost

Although ideally internal and external stakeholders will have a positive relationship with the employee association, there is also the potential for a negative relationship. For instance, the human resources department may think its territory is being usurped by the employee association, or the commercial establishments might argue that a new facility for the employee association will ruin their business. Volunteers may be more interested in a particular cause, or in obtaining a particular form of experience, rather than meeting the goals of the employee association. Although it is not always possible to achieve, a positive relationship with each stakeholder is something to be strived for.

CLASSIFYING THE STAKEHOLDERS

The power of stakeholders varies considerably, therefore each known stakeholder can be classified by the amount of power it possesses. Stakeholders can usually be easily classified into one of three groups, each with its own characteristics and share of power: gatekeepers, effectors, and activists. Determining the role a particular stakeholder has in the association may be a process of trial and error at first, but patterns quickly develop that indicate who has what role. In the following sections, each role is examined in detail and then an example showing how the three roles are interrelated is presented.

Gatekeepers

The gatekeepers are the most powerful of all the stakeholders, and they greatly influence decisions made by the employee association. Some employee associations have only one gatekeeper, while others may have two or three. If the gatekeeper approves of the program, normally all other stakeholders will accept the program even though they may not be in favor of it. However, if the gatekeeper does not approve of the program, it is almost certain to fail. Gatekeepers are sometimes hard to identify, because even though their opinions can make or break a program, they may not be personally involved in any employee association program. Theoretically, any of the stakeholders can potentially be the gatekeeper.

Effectors

Effectors are the doers, those groups who actually carry out the work of the employee association. They make the day-to-day decisions, and unlike the gatekeepers, they are very personally involved in the association. These groups usually have only moderate power and can never overrule the gatekeeper, but if they lack the technical skills necessary to carry out a project, then the project will probably fail. Any of the stakeholders can potentially be effectors, however, there are only a few effectors in each employee association.

Activists

The activists comprise the majority of the stakeholders. Although there are many of them, they only have power when they present a united front. Because the activists are so diverse, they are rarely united and therefore rarely a force equal to the gatekeepers and effectors. The best the activists can do to create change, is to make their wants and needs known, and then let the gatekeepers and effectors decide on the best way to meet them.

How the Three Roles Interrelate

These three roles may seem complicated, but each is seen clearly in the following example: A building proposal for a new gymnasium at Company X Employee Association passes the administrative board only after it has received the blessing of the city's town council; in this case, the city council is the gatekeeper. The company contract office and the employee association administrator have assumed the role of effectors, because they are the ones who will actually carry out the project. The environmental group, the marketing department, and all of the other stakeholders may have ideas about how the building project should be completed, but will have no power to implement any of their ideas, and therefore they are classified as activists.

The classification system is relatively permanent, but there are fluctuations of stakeholders between categories. If the city council is the gatekeeper at Company X Employee Association today, it is likely to be next week also. However, although the city council may be the gatekeeper at Company X Employee Association, at Company Y, the human resource department may be the gatekeeper and the city council may only be an activist. Understanding stakeholders' roles in an association as well as in neighboring associations is helpful in forming working relationships with stakeholders and other employee associations.

SOLICITING INPUT

Once the stakeholders have been identified and their share of power has been determined, it is necessary to seek the input of the effectors and the gatekeepers and, if possible, all remaining stakeholders. Stakeholder input should be genuinely sought and accepted, and should not be simply a token attempt to appease an interest group.

Informing, consulting, and delegating are three distinct forms of stakeholder involvement. Informing is basically telling the stakeholders about decisions that have already been made;

consulting is asking the stakeholders for opinions on decisions that have yet to be made; and delegating is providing the guidelines for decision making and then asking stakeholders to make decisions.   Depending on the situation, one type of stakeholder involvement will be superior to the other two, but on the whole, whenever an important decision can be delegated to a responsible group of representative key stakeholders, it is probably the best choice.

COMBINING STAKEHOLDERS' INTERESTS

The employee association administrator is faced with the task of combining all of the stakeholders' interests into one employee association strategy.   This is a tough assignment, because stakeholders may be opposed to each others' views. For instance, the physical fitness council may want the agency to build a gym, while the environmental group strongly opposes the employee association adding more buildings.   Delicate compromising and firm leadership skills are needed to keep the majority of stakeholders happy.

EXPLAINING THE DECISION

Although it is impossible to keep all of the stakeholders happy all of the time, the employee association can keep most stakeholders at least minimally satisfied by spelling out policies, explaining decisions, and providing rationales for decisions. However, although the stakeholders may be minimally satisfied with the decisions and policies, they will probably continue to subtly pressure the employee association to adopt views closer to their own.

It is important to explain decisions in the most appropriate way to each stakeholder. For instance, if the decision is made to build a new gym facility, the contract office may need to see blueprints and technical contracts, while the environmental group may need only a memorandum explaining why the balance tilted in favor of building instead of on preserving farmland.   Stakeholders are generally interested only in information that pertains to them, and providing them with excessive detail can be as frustrating to them as not providing enough detail.

CHAPTER SUMMARY

Stakeholders may focus on only one part of the employee association — the part that affects them — and from this they tend to form their entire impression of the employee association.

Stakeholders may be difficult at times, but without them the employee association would cease to exist. The employee association-stakeholder relationship benefits not only the stakeholder but also the employee association. Working with stakeholders costs time along with financial and human resources, but the dividends received from this investment are worth the price; after all, it is an "employee" association.

8

Working With Programmers

Although an administrator may not be directly responsible for the operation of a program, he must remember that each program is an investment, and that it can either produce great dividends, or cost the association dearly. Leaders of the employee association must constantly ask themselves if the results achieved are worth the time and money spent, since there are alternative programs and facilities that could be funded.

Programming and facility management go hand in hand. It is usually through programming that the association actually reaches the employee, but if the proper facilities or supplies are not available for a particular program, then the program cannot fully accomplish its goals. To receive maximum benefit from both facilities and programs, the facility administrator must see the employee association through the eyes of a programmer as well as the facility director. To do this, both the steps of programming and the benefits of programming must be understood.

STEPS IN PROGRAMMING

Five aspects are involved in programming, regardless of whether the program is a film, lecture, play, craft, or recreational activity. These aspects are (1) writing goals, objectives, and plans of action, (2) preparing for the activity, (3) implementing the activity, (4) immediately following up on the activity, and (5) a delayed follow-up activity.

Writing Goals, Objectives, and Plans of Action

Programmers and facility administrators generally begin at the same point - writing goals. In fact, their goals are very similar, usually stressing the desire to help employees reach their physical, emotional, mental, and social potentials. From then on, though, they differ considerably, because each has different objectives to meet these goals, and therefore each develops completely different plans of action.

Programming goals specify whom the desired audience is, and what the audience will be able to do as a result of the program. The objectives would indicate the steps to be taken to meet the goals, and the plans of action state how these objectives will be implemented.

Preparing for the Activity

In preparing for an activity, the programmer will generally ask the following questions. If any of the answers is no, then the programmer may drop the idea in favor of a different one:

1.   Does the activity meet the objective? If the objective is to meet the person's physical needs, then the activity must meet this need. It may meet more needs than the objective states, but it must meet those needs at a minimum.

2.   Do the likely participants have the ability to succeed in the activity? If they do not have the skills needed to succeed, is it worthwhile for them to learn them?

3.   Do the likely participants have an interest in the activity? On rare occasions the "try it, you'll like it" approach may be used, but planned programs should be geared to the interests of the likely participants.

4.   Are there adequate resources — time, supplies, facilities, money, and personnel — to carry out the activity at a satisfactory level?

5.   Is the activity too risky for participants? Assuming that

risks are controlled to the maximum extent that they can be, is the activity dangerous? Most activities involve some degree of risk, but for some it may be too high.

6.   Is this the best activity possible considering the resources available for meeting the objective? If not, should this be a back-up activity and another activity be implemented instead?

Planning the activity also includes obtaining supplies and equipment, finding qualified personnel, locating adequate facilities, and rearranging facilities to accommodate the activity. Those facility administrators who understand programming can offer help on how to arrange rooms, and which rooms are best for the activity. At this stage, programmers are also busy learning about the activity — its characteristics, its rules, and its procedures — and are trying to organize the activity into a detailed schedule. The more organized the programmer is, the more likely the facility administrator will provide assistance. A checklist of program planning steps utilized by one employee association is found in Appendix J.

Every program cannot be offered within each employee association facility. However, in planning outdoor programs and trips to neighboring communities, programmers may turn to the employee association administrator for help in selecting equipment, transportation, and lodging. Employee associations also often rent space and materials rather than purchase their own.

Implementing the Activity

Although all programmers work differently, their program organization usually follows the same basic pattern. First they secure the participants' attention with a verbal welcome or by rising from a chair or motioning for silence. They then position the participants into the desired formation, such as a circle, semi-circle, lines, rows, or clusters, and then position themselves in relation to the participants. Next they provide directions and supervise the activity, occasionally helping participants who are unclear about the directions.

Immediate Follow-up of the Activity

When the activity is winding down, the programmer will usually call all participants together to review and to answer any questions. The programmer may also obtain verbal and written feedback about the program that can be used to evaluate the program. By using these final moments as a summary time, the programmer can help participants to understand how to utilize the skills learned or developed by the program in everyday life to provide them with a richer, more meaningful life.

Immediate follow-up is generally considered as any follow-up done within an hour of the activity. Besides obtaining feedback and providing a summary, this time also includes putting away supplies, putting the facility back in its original condition, talking individually with any participants about the activity as they exit, and making evaluative notes.

Post Follow-up of the Activity

The purpose of the post follow-up is to determine if the activity was successful in meeting its objectives. This information is usually gathered from one of three sources: unbiased records, interviews, and observations. If the objective was that all overweight participants lose five pounds, the scale would provide an unbiased record. If the objective was for participants to become more attached to their mates, they can be asked if this was the outcome in an informal interview. If the objective was that the participants become "team players" at work, then observations by a supervisor would be an important source of information. Other sources of data can also be used, and it is important to choose the most reliable source of data to determine whether the program goals were met.

The post follow-up is also a time of writing thank you letters, attending to evaluation forms, and preparing for planning the next activity.

BENEFITS OF PROGRAMMING TO
AN EMPLOYEE ASSOCIATION

Although programming does have disadvantages, such as creating extra work for maintenance personnel, programming is essential to the employee association. Regardless of whether the programming is performed by paid professionals or merely volunteers, it produces results that can be grouped into three categories: financial, status, and goal success.

Financial

Although programs can be expensive — consuming supplies, energy, and staff time - they tend to produce revenue. This increase in revenue seldom comes from the program itself, but instead can be found in the increased traffic flow at the employee store, the snack bar, and vending machines.

Status

If quality programs are housed within the employee association's facilities, people become attached to the facility and hold both the facility and the employee association in high esteem because they associate them with the program. Programming may be an inconvenience at times, but it is a major foundation of public relations for the employee association and its building facility.

Goal Success

Programming is one of the best ways to ensure that the purpose and goals of the employee association are met. Although there are other ways to meet these goals, structured programming has an excellent record of success. Good programs are not ends in themselves, but are the means of reaching the ends (the goals of the employee association) that both the facility administrator and the facility programmer want the employee association to reach.

CHAPTER SUMMARY

Cooperation between facility administrators and facility programmers is necessary if the employee association is to be successful in meeting its goals. A program can not function without adequate support from facility administrators, and a facility cannot reach its potential without quality programming.

Operating Facilities

As the root of the word implies, a "facility" is something that facilitates. By using various facilities and physical space, employees are more able to experience the programs offered.

Most facilities are taken for granted. For instance, you have probably walked past a painting or a vending machine in the hallway every day since you first came to work and yet have probably not even realized that they are part of the physical space within the employee association and that therefore they can be modified to better meet the needs of the employees.

In order to enhance the quality of life for the employees, the employee association may find itself in control of a wide range of facilities. Facilities are not essential for an employee association to exist, but striving to receive as much mileage as possible from each facility which one has is an obligation of all employee association administrators. The following are brief summaries concerning the administration of common areas of operation; some employee associations will have nearly all of these while others will have relatively few.

Art Exhibits

From paintings on the wall to display cases for jewelry, art can be exhibited for all to experience. Art works can range from classical masterpieces to drawings by the employee association's art hobby club. If theft or accidental mutilation are not a threat, the art can be displayed anywhere that is convenient. If theft or mutilation is a likelihood, then locked display cases, an alarm system, or a security guard to watch an exhibit room may be necessary.

Arts and Crafts Center

The arts and crafts center is an area complete with tools and equipment necessary for constructing crafts. The equipment needed depends on the wants and needs of the employees, and crafts may include needlepoint, quilting, painting, woodworking and more. Besides providing employees with a sense of accomplishment, the arts and crafts center often saves employees expenses. For instance, rather than buying a table saw, an employee might simply pay an annual fee and use the association's saw in the crafts room. Classes to teach new skills/further skills are often available; these may be taught by volunteer or paid instructors. Often a full-time person is hired to supervise the instructional programs and maintain the arts and crafts center.

Box Office

Box office operations can range from a cigar box of tickets offered in a secretary's desk to a full-fledged ticket center. The tickets offered can cover a wide range of events, including amusement parks, movies, ice skating, live theater, professional sporting events, concerts, the circus, and the company softball tournament. The box office is generally able to purchase tickets at a lower price because it purchases in bulk. Even with a service charge to cover operating expenses, the box office price may still be far below that of any other source available to the employee.

Most ticket outlets will gladly aid the employee association's box office, because it saves the ticket outlet valuable time on the night of the show by having tickets presold. The box office also promotes the show and often guarantees a specific number of sold seats. If an employee association is unable to staff a full-time box office, but ticket sales are too complex to be handled by a volunteer, this function can be assumed by the business office, the information desk, or the employee store.

Break Area/Lounge/Television Room

The break area, the lounge, and the television room are often one and the same, or they may be three different rooms. A lounge generally has tables, chairs, a sofa, and a particular atmosphere - there are music lounges, reading lounges, gossip lounges, art lounges, and television lounges. Because quiet is desired and there may be a possibility of theft, attendants are often stationed in the art and reading lounges.

Lounges are generally open to all employees, family members, and guests. Break areas tend to be more exclusive, often open only to people currently on duty in areas operated by the employee association. The break lounge is often personalized to the tastes of those to whom it caters.

The television room is generally open to anyone allowed on company grounds. More elaborate employee association facilities have several television lounges; often there is one for each of the major broadcasting and cable networks.

Bulletin Boards

Bulletin boards can be a teaching tool, as well as a quality decoration, depicting the benefits offered by the employee association. They can also be a source of information on upcoming events, and for people to contact others for traveling, playing tennis, or renting property. To avoid the bulletin board becoming a cluttered mess, clear policies about who may post, what may be posted, and who must approve of the posting should be established and enforced. If theft and mutilation are potential problems, announcements can be placed behind a glass casing.

Display Cases

Display cases can serve many different functions. The function of displaying and protecting art has already been discussed. Another possible function is to advertise upcoming employee association events, such as the annual picnic. They can also be used to promote awareness of the services provided by the employee association, such as the employee store. Not only are display cases functional, they can also be used decoratively to lend atmosphere to a room. By changing display cases every few weeks, life is added to an otherwise drab area.

Business Office

The business office usually handles all accounting, budgeting, purchasing, and financial counseling done by the employee association, and may be staffed by one person, a small group, or company personnel. The business office is sometimes directly responsible for business operations such as the snack bar, the employee store, the newsletter, and the travel center. This office should be staffed by a person/people who are not only business-wise but who also understand programming. To promote harmony between business and programming, the employee association administrator may want to hold weekly staff meetings with program leaders and representatives from the business office.

Check Cashing/Branch Banking

Rather than require employees go to a local bank to cash their paycheck, the employee association will often cash the check or will have a branch of a local bank set up an office for this purpose.

When the employee association accepts direct responsibility for check cashing, the checks are usually cashed in one of four settings:

1. the employee association business office
2. the information desk
3. the employee store
4. the box office

Sometimes commercial banks are anxious to lease space from an employee association. These banks can install automatic teller machines that are open twenty-four hours a day. On payday, the bank may send personnel to the branch facility to handle the volume of business, and any problems that might arise.

Director's Suite

This is usually the brain of the employee association. It is often located in an out-of-the-way corner close to the information desk. If the intercom system is not headquartered at the information desk, it is usually located in this office for easy paging and for control of the airwaves. The director's suite is usually composed of a minimum of four rooms: the director's office, where the director meets with his guests; the secretaries' office, where the computers and intercom are located; a supply room, where various secretarial machines and supplies are kept; and a reception room, where guests are greeted when coming to see the director.

Employee Store

The employee store can be as simple as a supply of jackets with employee association logos for sale to a complete store carrying general merchandise.

Running an employee store is much like running an employee association, because policies, accounting procedures, and a hierarchy must be determined. The employee store is explained in detail in the Sobanski article found in the suggested supplemental readings at the end of this book.

Fitness Center

The fitness center consists of a combination of gyms, classrooms, and health testing equipment. The fitness center gyms are often used for organized exercise programs. Nutrition, cardiopulmonary resuscitation (CPR), and smoking cessation are just a few of the topics covered in the classrooms. The health testing equipment measures endurance, heart recovery rate, and physical condition. Professionals may be available to administer health check-ups. The fitness center is distinguished from a typical gym, because the center is much more structured.

Games Area

A games area may contain electronic video games, or more traditional games, such as ping-pong, billiards, and chess. A games area may require a full-time person to handle and maintain equipment, settle disputes, and enforce gameroom rules. If such an attendant is necessary, all games should be offered within the same room so the expense of hiring attendants is kept to a minimum. Even with the expense of an attendant's wages, the gameroom can be a major money maker, particularly if the video games are well utilized.

Gymnasium

The gymnasium and the multipurpose room are sometimes one and the same. The gym can be used for a variety of activities, limited only by the imagination. It can become a basketball court, a volleyball court, a shuffleboard court, and the list goes on. These can be separate courts, or in smaller facilities, they can overlap one another. Masking tape can be used to create temporary lines if permanent lines are undesired.

Facilities within a gymnasium can be very elaborate or very basic. Within the gym, one might find mats for wrestling, karate, and other contact sports; an equipment check-out desk; hot tubs, saunas, and whirlpools; a locker room; a lounge; a refreshment area; a rifle range; a running track; squash rooms; a swimming pool; weights; and classrooms for instruction in aerobics or dance.

Not all athletic facilities are in the gym. Common outdoor facilities include the nature trail, baseball diamond, football field, the soccer field, outdoor track, and park lawn. As with indoor facilities, many outdoor facilities can have dual purposes: the same turf that comprises the baseball diamond may also be used for the football field, soccer field, or rugby field by painting different lines on it.

A gym is sometimes a separate facility, but it also may be part of the office complex. Gyms tend to be warmer and more humid than the rest of the office complex, so adequate ventilation is important. Although having facilities close by requires careful monitoring, the convenience of having them so near often encourages greater use.

Information Desk

The information desk is the hub around which the rest of the facility turns. The information desk should be easy to locate and is therefore immediately inside the main entrance or at the center of a "Y." If there is no room for an information desk within a facility, the information clerk may double as a cashier at either the snack bar or the employee store, or as the receptionist for the director of the employee association. Because of the volume of telephone calls and traffic, the information clerk is often only a figurehead when doubling at another station.

If possible, the information desk should have maps of both the company and the community to distribute to all visitors. It should also have drawings of the floor plans of each building guests might want to visit. In addition to having written information handy as a reference, the desk clerk should be able to field most questions and know where to find answers to those that are unknown. Although the information imparted is often about the employee association and its facilities, the desk clerk should be able to give basic information regarding any aspect of the company, or of the community as a whole.

The information desk also acts as a security arm. If a suspicious person enters the facility, the information desk clerk should quickly notify supervisors or the security department. Many people will call the information desk in times of panic, asking how to handle a situation, to get an emergency phone number, or to find a particular individual. The desk clerk must be efficient but calm when handling crises.

The information desk clerk is often the first person visitors, new employees, and prospective employees see upon entering company grounds. The initial encounter with the desk clerk can set the tone for the rest of their stay with the company and the employee association. All three functions are important - giving information, offering security, and providing public relations - and the successful information desk will cover them all.

Janitorial/Maintenance Center

Most employee associations have enough facilities to merit at least a part-time janitor and maintenance crew to keep the facilities cleaned and repaired. These personnel need a headquarters to organize themselves and to store supplies. Even if the company provides maintenance personnel, there are still benefits from a janitorial/maintenance center for the association. In addition to maintenance tasks, night-time janitorial personnel often serve the purpose of security guards.

The janitorial center should contain all the supplies that the maintenance personnel needs to perform their assigned tasks. The center may also serve as a break room for maintenance personnel, but care must be taken to see that the facility is not abused, since personnel who are constantly on break do not clean hallways, repair facilities, or scare away prospective thieves.

Lockers/Locker rooms

People often bring personal items to work such as gym clothes, books, and purses — items they don't want to carry around, but that they don't want to leave on a counter. Lockers allow people to safely store items and maintain a neat, safe environment. There are three basic types of lockers, each with its own advantages and disadvantages: bring-your-own-lock lockers, coin-return lockers, and coin-operated lockers.

The bring-your-own-lock lockers are free of charge to employees, and generally create good will. However, some employees leave locks on indefinitely, rendering the locker useless to other employees. Another problem is they don't provide revenue, even though it costs to maintain them.

Coin-return lockers are also generally appreciated by the employee, because they get the deposit back. Because one master key unlocks all of the lockers, administrators do not have to worry about cutting or picking locks. With this type of locker, keys are occasionally lost, a problem seldom encountered with the bring-your-own-lock lockers. By charging a substantial lost-key fee, patrons become more conscious of their keys. The operation of coin-return lockers requires correct change for deposit, which may be inconvenient.

Coin-operated lockers pay for themselves. Although users may grumble about the daily fee, nonusers can argue that it is only fair that the users pay the maintenance fee since they are the ones using the lockers. Like coin-return lockers, coin-operated lockers have the advantage of having a master key, and the disadvantage of the possibility of individual keys becoming lost.

Many gym facilities have three different locker rooms - men's, women's, and visitors'. The men's and women's are for all members of the employee association, and they may leave their items in their lockers as long as they are members. The visiting locker room is for teams from other companies, and no items may be left in it overnight. Having a third lockerroom not only guarantees facilities to guest teams, but it also cuts down on theft by guests.

Multipurpose Room

The multipurpose room is exactly that - a room with many purposes. For instance, this might be a multipurpose room's weekly booking: on Saturday children will be shown a series of cartoon movies, on Sunday the baseball club will have a recognition banquet, on Monday the drama club will rehearse a play, on Tuesday the square dance club will practice, on Wednesday a guest speaker will lecture on wellness, and on Thursday the room will be segmented so each club has a place to meet.

Because of its flexibility, the multipurpose room is usually in great demand. If this is the case, its use should be scheduled so that no conflicts arise. The scheduling of this room is usually handled by either the employee association business office, or the employee association information desk.

Newsletter Office

The employee association's newsletter is an important tool for both public relations and spreading news. Often the newsletter staff, whether volunteer or full-time paid workers, will have its own office and printing equipment. If there is no office dedicated exclusively to the newsletter, space within the employee association's business office should be set aside for people to fill out information forms about items that they would like to see printed in the newsletter.

Nursery/Day Care Centers

This room and its adjoining outdoor lot should be stimulating to children's imaginations. This stimulus can come from murals painted on the walls, or simple toys such as sandboxes. Furniture should be child-size; this includes tables, chairs, and even the restroom facilities. Safety is a major concern, and cribs and toys should be chosen carefully to prevent strangulation, choking, and cuts.

Outdoor Patio

Stepping outside during break can be very relaxing. In general, the outdoor terrace is located near the snack bar or vending machine, so employees can enjoy the outdoors while refreshing with a soda. Some programming, such as a "brown bag luncheon discussion group," may occasionally take place on the terrace.

Photography Club Darkroom

Black and white photography is a hobby many enjoy, and having a darkroom sponsored by the employee association can save employees substantial commercial developing costs. The photography club is one of the oldest of the employee association special interest clubs and it usually assumes some responsibility for running the darkroom. Use of the darkroom may or may not be limited to members of the photography club.

Photography is relatively expensive, and for the novice, relatively complicated. Because of this, the darkroom may be staffed during certain hours, such as 6-8 a.m. and 7-10 p.m., with quality instructors who will answer questions and demonstrate processes. Formal workshops on how to compose pictures and how to use a darkroom may be offered from time to time.

The employee association will also have to decide on the type of materials to keep on hand in the lab. For instance, does it want to bulk order film, saving employees lots of money, but creating financial headaches for itself, or does it want employees to buy their own film? Since the photography club members are among the biggest users of the lab facility, the director may want to attend a photography club meeting to explain the pros and cons involved in each issue, and obtain feedback from the club members, prior to making decisions about the darkroom.

Post Office Branch

Most employee association facilities have a designated box where mail can be dropped off for pickup by United States postal workers. Most employee associations also sell stamps at the information desk, the business office, and/or the employee store; those employee associations that do not want their employees to bother with selling stamps usually offer vending machines where stamps can be purchased.

Security Office

In many employee associations, security is a function rather than a position; however, the duties of the security guard are fulfilled even in the facilities which have no security guard per se. Someone prosecutes shoplifters, handles fistfights, and assumes responsibility for the building at night. The information desk, the facility administrator, and the night janitor often supplement the security office.

The decrease of theft/vandalism often pays the salary of the security personnel and the feeling of safety among patrons is good for public relations. Besides handling crimes, the security office aids in medical emergencies, sponsors programs with a safety theme, and supervises employee association parties. This office is often responsible for keeping a log, and this log can help the facility director pinpoint areas where trouble constantly occurs so that corrections can be made. For instance, if prowlers have been spotted near a rear door several times, the director may install a new outside light by the door. Having a security office helps employees know where to turn in case of an emergency. The security office is often located near the information desk and the facility administrator's office.

Shower Room/Laundry Room

The shower room is usually connected to a restroom. Shower room facilities may be very simple– with only hooks for hanging clothes, a bench, and a mirror, to very lavish– with brightly painted lockers for clothes, benches, chairs, sofas, and a full-length mirror as well as bathroom mirrors. The more lavish facilities may even have a towel room attendant. The shower room is among the most dangerous of all of the employee association facilities because it is so slippery, and safety precautions should be emphasized.

The laundry room, often consisting of only one washer and a dryer, is often a part of the shower room. To avoid abuse and theft, this area is off limits to most employees, and a designated person will collect and wash sport team uniforms, snack bar uniforms, and other employee association laundry.

Snack Bar

Snack bar prices generally must be a few cents more than vending machine prices because of the cashier's salary, but many people appreciate the human contact and will gladly pay the extra pennies. Because many of the same items can be found in a vending machine, it is courteous service, rather than menu items that make or break snack bars.

Due consideration should be given to the decorating and layout of the snack bar. To promote atmosphere, many snack bars have a theme, such as a reflection of the company, the 1950s, railroads, or the farm, and all decorating centers around this theme. The layout of the snack bar should reflect how people enter and exit, and how each foot of the store can be maximized.

Snack bar advertising is unique, because instead of relying solely on sights and sound, the snack bar can also rely on smell. Who can resist the smell of hot buttered popcorn? Pleasant smelling products attract people and sell other products in the store.

Travel Center

Travel center services can range from leaflets from a travel agency, to selling space on busses and planes, to a travel agent actually being on the premises. The travel center is usually managed by the information desk, a professional travel agent, or the program office.

Each source of management has its own characteristics. If the information desk handles the travel center, it may consist of maps and details of group trips. If a professional travel agent handles the travel center, it will actually book both individual and group trips. If the program office handles the travel center, it will usually book group trips but only offer information about individual opportunities to travel. Deciding who manages the travel center often determines how the travel center will function.

Vending Area

Due to both increasing technology and rising labor costs, vending machines are becoming more and more common. Vending machines can market almost any product, including bank money, soda pop, pencils, pretzels, cigarettes, and condoms. Whenever a vending machine is utilized, a dollar-bill-changer should be nearby to provide necessary coinage; this bill-changer can be either a machine or a clerk. Although vending machines are less personal than a cashier, they can operate twenty-four hours a day at a fraction of the cost. The location of the vending room varies, but often it is in a central location near the outdoor patio.

A vending area may also be located near the gym, for employees who work up a thirst and hunger exercising. For nutrition's sake, health foods are normally offered as well as favorites such as soda pop and candy. To increase the educational value of a refreshment station, one may post the number of calories and the amount of vitamins found in each food item. Color-coding selection buttons so patrons know which food group the item comes from may be appreciated.

Vending machines are normally grouped by purpose. For instance, the candy, chip, and soda machines are often side by side. Some facilities choose to have all vending machines located in one area, such as a lounge or hallway, while others prefer having machines scattered throughout the building. The proponents of each of the distribution techniques argue "convenience". "Why walk all over the building looking for a chip machine when all vending machines can be in one central location" vs. "Why walk to a central location when I can have the soda machine right outside my door?" Both of these arguments have merit, and a compromise may be having most vending machines placed in a central location and having duplicates of popular machines located in other areas. By having most vending machines grouped together, and having an outdoor exit, access to them can be gained after hours while the rest of the building is locked for security. An outside entrance to the vending room also makes it easy for the distributor to restock them.

Vending machines are a significant contributor to an employee association's income, ranking in the top three with company contributions and employee dues. For this reason, vending machines are an important part of an employee association, and keeping them restocked is only one of many responsibilities they bring.

Meeting Rooms

Most employee associations have four different types of meeting rooms: the board room, the conference room, the club room, and office space. Each type has its own characteristics and its own purpose.

Boardroom

The board room is generally a room with a long table surrounded by chairs. Normally it is used for official business meetings when little creativity is sought. The board room is normally decorated to convey a formal atmosphere,

Conference room

The conference room has furniture that can be easily arranged in any desired fashion. The conference room generally has a less formal atmosphere than the board room. If one wants to create an even less formal atmosphere, the meeting might be held in the lounge.

Club room

The club room is space set aside for one particular club of the employee association. For instance, the model railroading club might need its own room to set up a display. A chapel for religious groups would be relatively permanent, as would be a darkroom for the photography club. In general, any employee association subgroup that is greatly inconvenienced in setting up each time deserves its own room. However, because space is limited, and competition is great, the director will often allocate space on an annual basis.

Office space

Office space, like club rooms, is generally assigned on a yearly basis. Office space is normally allowed for official business personnel such as the director of volunteers, and the employee association business manager. If space remains for office use, each club may state its case as to why it needs office space more than other clubs within the employee association. For instance, the softball club coaches may want space to store scorebooks and scouting reports, to interview prospective team members, and to discuss strategy.

Section Summary

The facilities mentioned in this section are among the most commonly found employee association facilities, but numerous others also exist. None of the facilities mentioned is a necessity and yet each offers a new dimension to the employee association. The operation of these facilities is only a means to an end — to help the employees reach their full potential physically, emotionally, mentally, and socially — but to reach that end the facilities must be operated skillfully.

Entire books have been written describing the operation of each of these facilities. In general, one should know the function of each facility, possible safety hazards of each and how to minimize them, general care and maintenance techniques, the potential of each facility, and how to better maximize that potential. Manuals describing the detailed operations of these facilities should be read, and their contents applied to the goals and mission of the employee association.

CONCLUSION

Although no two employee associations are exactly alike, all exist for the same purpose: to help the employee have a more fulfilled life. This book has attempted to give a broad overview of the history, philosophy, and administration of the employee association. Employee associations are complex, but it is the aim of this broad perspective to show how all of the parts of the employee association fit together into a whole. This book is not meant to be the final word on employee association administration — it is rather a foundation upon which further understanding can be built.

Suggested Supplementary References

Bannon, J. J. 1985. *Leisure Resources: Its Comprehensive Planning*. Champaign, IL: Management Learning Laboratories. A good overview of the purpose and strategies of comprehensive planning. It also contains a glossary of planning terms and an appendix of working forms including a sample survey.

Busser, J. in print. *Adult Programming for Employee Services*. Champaign, IL: Sagamore Publishing. An in-depth book about programming written especially for the employee association.

Brattain, W. E. 1981. *The Administration of College Unions and Campus Activities*. Tennessee, IL: self-published. An excellent book on facility management. Although it focuses on student unions and campus recreation, much of what is said can easily be translated to the employee association. The book also contains a glossary of food service terms and of budget terms as well as an appendix of scheduling sheets and other working forms.

Deppe, T. R. 1983. *Management Strategies in Financing Parks and Recreation*. New York: John Wiley & Sons. An elementary introduction to the purpose and techniques of budgeting. It contains a glossary of financial terms and working financial forms.

Fink, A., and J. Kosecof. 1978. *An Evaluation Primer*. Beverly Hills, CA: Sage Publications. A step-by-step outline of the evaluation process.

Frenkel, T., 1982. "Planning the Great Escape." *Employee Services Management*, 25, September, pp. 14-19. This article about travel is written in an easy to follow question and answer style and contains a useful glossary of travel terms.

Heidrich, K., in print. *Volunteers in Employee Service Organizations*. Champaign, IL: Sagamore Publishing. This book provides an in-depth examination of volunteers and of volunteering specifically written for employee associations.

Henderson, J., and B. L. Marple. 1988. *Zingers*. New York: American Management Association. The authors have devoted one chapter to profound insight into obtaining a meaningful internship.

Henke, E. 1985. *Introduction to Nonprofit Organizational Accounting*. Boston, MA: Kent Publishing Company. This is a technical book full of examples; although this book is not particularly directly meaningful to most employees, it is almost indispensable to the financial chair.

Kelsey, C., and H. Gray. 1985. *Master Plan Process for Parks and Recreation*. Reston, VA: American Alliance for Health, Physical Education, Recreation, and Dance. An excellent step by step book describing the comprehensive planning process.

Klarreick, P. R. 1987. "Health Promotion in the Workplace: A Historical Perspective." Samuel H. Klarreick, Ed. In *Health and Fitness in the Workplace*. New York: Praeger. This article is an indepth examination of how corporate fitness evolved.

Klarreich, S., Ed. 1987. *Health and Fitness in the Workplace*. New York: Praeger. This book is a series of articles about employee recreation and wellness.

Kraus, R. 1984. *Recreation and Leisure in Modern Society*. Third edition. Dallas: Scott, Foresman, and Company. A good overall view of recreation in general, showing how employee recreation integrates with other forms of recreation.

"Large/Small-Company Programs: Composed of the Same Elements," 1986. *Employee Services Management*, 31, Oct. The articles within this edition of the magazine profile a large employee association and a small employee association; although the articles themselves are not exceptional, the concept of the comparison is.

"Liability waivers." *Executive Update,* January, 1989, p. 35. This article notes that liabilitiy waivers are beginning to be recognized by judges as valid.

McCaffery, R. M. 1983. *Managing the Employee Benefits Program*. New York: American Management Associations. This book contains a history of employee benefits and a glossary of legal terms.

McCormack, M. 1984. *What They Don't Teach You at Harvard Business School.* Toronto: Bantam Books. An excellent how-to book on how to maximize the investment you have in yourself.

McVicca, M., and J. F. Craig. 1981. *Minding My Own Business: Entrepreneurial Women Share Their Secrets of Success.* New York: Richard Marek Publishers. Although this book is geared to women beginning a private business, many of the concepts can be applied to employee associations that are starting a company store, a snack bar, or any other facility.

O'Connell, B. 1976. *Effective Leadership in Voluntary Organization.s.* Chicago: Follett. This neatly organized book provides general tips of how to work with voluntary associations in general, particularly in the aspects of funding and volunteers.

Ribaric, R. F. 1982. "The Company Open House." *Employee Services Management,* 25, September, pp. 14-19. This article provides a step-by-step analysis of the how's and why's of hosting an open house.

Santa-Barbara, J. 1987. "Corporate Health and Corporate Culture". Samuel H. Klarreick, Ed. In *Health and Fitness in the Workplace.* New York: Praeger. An excellent article focusing on the philsophy of why a corporate recreation program is useful and why each corporate recreation program must be unique.

Shaw, M. E. 1981. *Group Dynamics: The Psychology of Small Group Behavior.* Third edition. New York: McGraw-Hill Book Company. The author incorporates numerous research studies into one book to aid the administrator in working with/being a part of groups.

Shephard, R. 1986. *Fitness and Health in Industry.* Paris: Karger. This technical book attempts to assemble numerous up the minute research studies about employee recreation into one understandable book.

Shinew, K. 1988. "Bottom-line Results for GE." *Fitness Management,* Nov./Dec., pp. 22-23. 28. An excellent article describing the findings of a research study of corporate fitness at General Electric.

Schultz, J. , in print. *Revenue Generation and Management*. Champaign, IL: Sagamore Publications. A detailed examination of revenue generation and fiscal management applied specifically to employee associations.

Sobanski, A. 1988. "The Company Store: A Spectrum of Items." *Employee Services Management*, 36, April, pp. 13-17. This article is a detailed examination of administrating the company store.

Tober, P. 1988. "Historical Perspective: The Evolution of Employee Services and Recreation." *Employee Services Management*, 31, February 1988, pp. 11-16. A historical overview of the past and a projection of the future of employee services. This article also contains several pictures from the early days of employee associations.

Webster, G. 1988. *The Law of Associations*. New York, Bender. Webster is a general counsel to the American Society of Association Executives and this indepth reference book is updated annually. This book is a basic problem solver for the technical aspects of associations.

Wilson, M. 1976. *The Effective Management of Volunteer Programs*. Boulder, CO: Volunteer Management Associates. Although the book focuses on volunteer management, it is an excellent introduction to management of all personnel.

Wolfe, T. 1985. *The Nonprofit Organization: An Operating Manual*. Englewood Cliffs, NJ: Prentice-Hall, Inc. Although this book claims that one must be incorporated to be a true nonprofit organization, much of its contents can be applied to all employee associations.

Yeomans, W. 1985. *1000 Things You Never Learned in Business School: How to Get Ahead of the Pack and Stay There*. New York: McGraw-Hill. This book and McCormack's are very similar but each has some unique material to stimulate a manager's thinking on how to get better results faster.

Appendix A
Organizational Chart

EMPLOYEE CLUB ORGANIZATION CHART

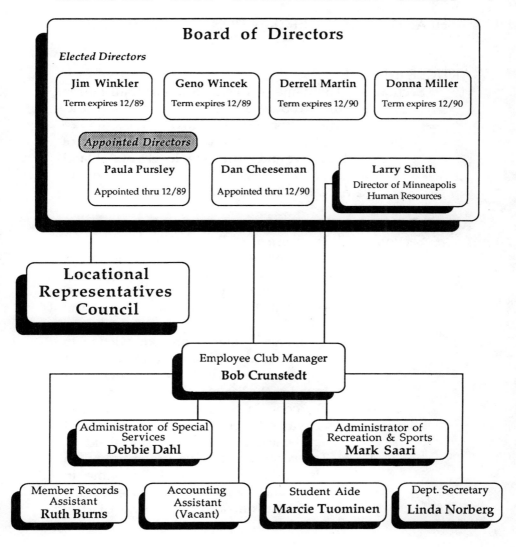

Revised 03/89

ASHLAND TRI-STATE RECREATION ASSOCIATION, INC
CONSTITUTION

ARTICLE I: NAME
> Section A. The name of this association shall be the Ashland Tri-State Recreation Association, Inc. (hereafter referred to as ATRA)
>
> Section B. It shall be a voluntary non-profit association of eligible active and retired employees of Ashland Oil, Inc, and its affiliated companies, located in the Ashland - Tri-State area.

ARTICLE II: PURPOSE
> Section A. The purpose of ATRA shall be:
>> 1.   To promote a spirit of fellowship, participation, and understanding among its members by encouraging a broad program of social, cultural, and physical activities.
>> 2.   To increase interest and knowledge in recreational activities and to offer opportunities which will enrich and broaden the life of every member.
>> 3.   To coordinate and assist members' activities by providing leadership and equitable distribution of financial aid.
>> 4.   To accept and receive property and to manage and dispose of the same in accordance with the purpose of the Constitution and By-Laws of the association.

ARTICLE III: MEMBERSHIP
> Section A. Individuals and immediate families of individuals who are regular co-op students and part-time employees, regularly working 30 or more hours per week and retired employees and their families shall be eligible for membership in the association.
>
> Section B. Membership dues shall be fixed by the By Laws.
>
> Section C. Membership in the association may be suspended, terminated, or reinstated for cause by a two-thirds vote of the Activities Council after a fair and impartial hearing.
>
> Section D. A member in good standing shall be entitled to:
> 1. A vote and to hold office.
> 2. Propose the creation of a new activity.
> 3. Admittance to all scheduled functions of the association subject to the individual requirement of the function.
> 4. Participation in all activities of the association, subject to the individual requirements of the activities.
> 5. The rights prescribed by the Constitution and By-Laws of the association.

ARTICLE IV: MANAGEMENT
> Section A. For the control and management of the association and its property there shall be an Activities Council and a Policy Committee.
>
> Section B. The Activities Council shall consist of
> 1. Four representatives elected by the members employed in the downtown Ashland area (District I).
> 2. Four representatives elected by the members employed in the Huntington-Catlettsburg area (District II).
> 3. For representatives elected by the members employed in the Greenup County area and other at large company locations designated by the Activities Council (District III).

The three (3) districts listed herein are subject to re-appointment by action of the Activities Council as set forth in the By-Laws.

Section C. Representatives shall be elected during the month of August and will take office September 1st.

Section D. Representatives shall be elected for a two-year term.

Section E. The Policy Committee shall consist of:

1. Four representatives appointed by Ashland Oil, Inc.
2. The Elected President of the Activities Council will also serve as a member of the Policy Committee.
3. The representatives of the Policy Committee shall serve as the Board of Directors.

ARTICLE V: EXECUTIVE OFFICERS

Section A. The officers of the association shall be a President, Vice President, Secretary, and Treasurer.

Section B. New Officers of the association shall be elected annually by the Activities Council from its elected representatives at the September meeting of each year.

ARTICLE VI: APPOINTMENTS

Section A. All appointments made by the President shall be subject to confirmation by a majority of the Activities Council.

ARTICLE VII: MEETINGS OF THE ACTIVITIES COUNCIL

Section A. The annual meeting for the election of officers shall be held by the Activities Council on the third Monday in September or as near thereto as circumstances will permit.

Section B. The Activities Council shall hold regular monthly meetings on or before the third Monday of each month.

Section C. Special meetings of the Activities Council may be called by the President or in his absence, the Vice President.

Section D. A majority of the elected representatives shall constitute a quorum of the Activities Council.

ARTICLE VIII: MEETINGS OF THE POLICY COMMITTEE

Section A. The Policy Committee shall hold meetings at its discretion and on call from the Recreation Coordinator.

Section B. A majority of the representatives of the Policy Committee shall constitute a quorum of the Policy Committee.

ARTICLE IX: VACANCIES

Section A. In the event of vacancy in any one of the executive offices the Activities Council shall elect a representative to the office for the unexpired term.

Section B. A vacancy on the Activities Council shall be filled at the discretion of and in the manner prescribed by the Activities Council.

Section C. A vacancy on the Policy Committee shall be filled by appointment of the Recreation Coordinator.

ARTICLE X: FUNDS

Section A. All ATRA membership dues, company contributions, and other ATRA income will be placed in ATRA's bank account or an appropriate ATRA fund as determined by the Policy Committee.

ARTICLE XI: AMENDMENTS

Section A. Amendments to this constitution shall require a majority recommendation of the Activities Council and of the Policy Committee and a favorable vote of a majority of the ballots cast by the general membership at the annual election.

Section B. Notice of such an election and the proposed amendment shall be posted upon the bulletin boards of the Company at least two weeks before the date of said election.

September 1978

Appendix B Pt. 2

BY-LAWS OF THE ASHLAND TRI-STATE
RECREATION ASSOCIATION, INC.
(A Non-Profit Organization)

ARTICLE I: PURPOSE

The Ashland Tri-State Recreation Association, Inc. (hereafter referred to as ATRA) is organized to be operated exclusively for the purpose of furthering the morale, social and physical welfare of the active and retired employees and families of Ashland Oil, Inc. and its affiliated companies located in the Ashland Tri-State area, through social and recreational activities.

ARTICLE II: Principal Office:

The principal office for the transaction of the business of the Ashland Tri-State Recreation Association, Inc., is hereby fixed and located at the Ashland Oil, Inc. Home Office Building, ground floor, 1409 Winchester Avenue, P.O. Box 391, Ashland, Kentucky, 41101.

ARTICLE III: MEMBERSHIP

Section A. Eligibility. Individuals and immediate families of individuals who are regular, co-op students and part-time employees, regularly working 30 or more hours per week and retired employees and their families shall be eligible for membership in the Association.

   1. Application forms and descriptive literature for ATRA membership will be given to eligible new employees by the Employment Department.
   2. Application forms and descriptive literature for currently employed and retired employees will be provided by the Association to be completed and submitted to the Recreation Coordinator for membership.

ARTICLE IV: MEMBERSHIP DISTRICTS

Section A. The Ashland/Tri-State area will be divided into three (3) Election Districts. A District shall correspond to a physical sub-division of the Company's work locations in the tri-state area. One month prior to Activity Council elections, the Council shall review District work locations and re-apportion them if necessary to assure that employee representation is as equal as possible between Districts. Retired employees shall be included in the District that includes their last work assignment.

Section B. Every year during the month of August, two (2) members from each District will be elected to serve a two (2) year term. After the association's first year, each District will be represented by four (4) members on the Activity Council with one-half of each District representatives up for election each year.During the association's first year of business two members of the Activities Council from each election District will be selected by the Activities Council to serve during the association's second year of business and two (2) members from each District will be elected by majority vote of the membership at-large to serve two years.

ARTICLE V: ELECTIONS (ACTIVITIES COUNCIL)

Section A. At least one month before the annual election of the Activities Council, the President shall appoint a Nominating Committee, none of whom shall be a member of the Activities Council. There should be at least one from each election District.

Section B. The Nominating Committee shall arrange for the nominations of all candidates for the Activities Council.

1. They shall prepare a list of at least twice the number of representatives to be elected.
2. They shall solicit nominations from the membership.
3. Written consent of the nominees shall be obtained in all cases.
4. Activities Council representatives may be nominated.
5. All those holding active membership shall be eligible for nomination.

Section C. The names of all eligible candidates shall be published at least two weeks before the election.

Section D. The Nominating Committee shall prepare ballots, distribute ballots, and conduct the election.

Section E. Newly elected Activities Council representatives shall assume responsibility for their duties on the first of September, following the election.

Section F. The retiring Activities Council representatives

may serve in an advisory capacity to their successors for a period of at least one month but not more then two months.

ARTICLE VI: DUTIES (ACTIVITIES COUNCIL)
Section A.
1. The duties of each and every representative of the Activities Council shall be as follows:
   a) To represent the association members of the District from which elected, having due regard for the best interests of the association as a whole.
   b) To present to the Activities Council all communications pertaining to the association and duly signed by a member in his/her election District.
   c) To attend all meetings of the Activities Council.
   d) To serve as chairperson, member or adviser to standing or special committees as assigned.
2. The authority of the Activities Council shall be as follows:
   a) To elect the executive officers of the association.
   b) To fill vacancies on the Activities Council.
   c) To recommend annual dues and charges for revenue-producing activities.
   d) To confirm or reject for cause, applications for membership.
   e) To suspend or expel any member of the association for cause, after a fair and impartial hearing.
   f) To suspend or terminate any activity of the association for cause after a fair and impartial hearing.
   g) To confirm or reject appointments made by the President.
   h) To approve additional activities and budgets.
   i) To approve an annual budget.
   j) To recommend the borrowing of money, and purchase or sale of ATRA real and personal property.
   k) To approve any apportionment of Districts and determine the number of representatives to be

Appendix B119

elected within each District.
l) To formulate the rules and regulates of the association and to attend to all business not otherwise specified herein.
Section B. Executive Officers (Activities Council)
1. The duties of the President shall be as follows:
a)To serve as official representative of the association and a member of the Policy Committee
b)To serve as chairperson of the Activities Council and preside at its meetings.
c) To call special meetings of the Activities Council.
d)To appoint committees and committee chairpersons.
e) To sign financial statements and reports submitted to the members.
f) To issue an annual report.
g)To assume the duties of the treasurer in his or her absence.
h)To see that the Constitution and By Laws are enforced.
2. The duties of the Vice-President shall be as follows:
a) To assume full responsibility for the duties of the President in his/her absence.
b) To assume responsibility for the coordination of Activities committees functions.
3. The duties of the Secretary shall be as follows:
a) To issue all calls for meetings under the direction of the President.
b)To call roll at all meetings and ascertain if a quorum is present.
c) To prepare and submit minutes of the meetings to all representatives of the Council and Policy Committee.
d) To perform all other duties incidental to the office not otherwise specified herein.
4. The duties of the Treasurer shall be as follows:
a) To assume responsibility for the coordination of the Finance Committee as appointed by the President.

b)To countersign all checks signed by the Recreation Coordinator.

c) To issue only checks countersigned by the Recreation Coordinator or, his/her designated representative.

d)To furnish a bond, the amount to be specified by the Activities Council, and paid for by the association.

e) To perform all the duties incidental to the office not otherwise specified herein.

ARTICLE VII: POLICY COMMITTEE (APPOINTED BY ASHLAND OIL, INC.)

Section A. The duties of each and every representative shall be as follows:

1. To serve as the Association's Board of Directors.
2. To insure that the activities shall not be harmful to the interest or reputation of the Company.
3. To act in a liaison capacity between the association and the Company.
4. To attend all meetings of the Policy Committee.

Section B. The authority of the Policy Committee shall be as follows:

1. To approve the annual dues and charges for revenue-producing activities.
2. To review and sanction the Council's approval of association activities and budgets.
3. To periodically audit the activities program accounts.
4. To approve the borrowing of money and the purchase and sale of ATRA real and personal property recommended by the Activities Council.
5. To exercise all other powers vested in the Board of Directors by the laws of the Commonwealth of Kentucky.

Section C. Recreation Coordinator (Provided by Ashland Oil, Inc.)

1. His/her duties shall be as follows:

a)To provide guidance and counsel to the representatives and committee chairperson.

b) To assist members in starting and organizing activities.

c) To assist all leagues and clubs with organization, by laws, elections, communications, etc.

d) To attend all meetings of the Activities Council as a non-voting participant.

e) To serve as advisor to the Policy Committee without voting privileges.

f) To act in a liaison capacity between the Activities Council and the Policy Committee.

g) To assume responsibility for the correspondence of the association.

h) To maintain a list of the names of all persons accepted for membership.

i) To initiate, process, and submit for approval recreation invoices and disbursements of checks to recreation leaders (league/club officers), officials, and suppliers.

j) To sign and present all association checks to be countersigned by the association treasurer.

k) To keep a detailed record of the finances of the association.

l) To provide the Activities Council and the Policy Committee with quarterly financial statements.

m) To furnish properly signed receipts for all monies received and obtain proper receipts for all monies expended.

n) To maintain records concerning membership, employee participation in recreation programs, operation costs, and other related information.

o) To coordinate association programs with community activities and available community facilities.

p) To select and obtain facilities and equipment for recreation activities and to recommend the need for association facilities where justified.

q) To prepare and obtain approval of association budgets and proposals.

r) To conduct surveys and prepare materials for, and work with the association's Activities Council.

s) To initiate and develop new recreation programs as directed by association members and the Activities Council.

t) To administer all recreational activities and events.

u) To develop and maintain a working knowledge of trends in recreation programs and facilities.

v) To perform all other duties incidental to the office not otherwise specified herein.

ARTICLE VIII: CLUBS, LEAGUES, ETC.

Section A. Purpose. In an effort to offer the best variety of leisure time activities to active and retired employees of Ashland Oil, Inc., the Ashland Tri-State Recreation Association (ATRA) encourages the employees, retirees, and their families to pursue their own special interests through group participation. Clubs, leagues, and other activities are given assistance to organize and function under the auspices of ATRA.

Section B. All ATRA sponsored activities which use stationery, decals, patches, etc., must include ATRA identification (LOGO) in their design.

Section C. Objectives. Activities are formed and maintained in response to interest by Ashland employees who have a common desire to participate in a particular leisure time pursuit. By organizing and affiliating with ATRA, activities derive assistance in financing, facilities, guidance, communication, organization, legal identity, and liability coverage.

Other benefits of ATRA are as follows:

1. To better develop individual goals in their recreation pursuits by an exchange of information, ideas, and skills.

2. To assist each member through collective efforts - material, education, etc.

3. To develop friendships through social and fellowship programming.

4. For satisfaction received from belonging to a group and helping others.

Section D. <u>Organization</u>. ATRA members who wish to form an ATRA-sponsored activity must provide the following to the ATRA Recreation Coordinator:

1. A roster of interested members (minimum 10 people)
2. A list of provisional officers.
3. Provisional by laws.

If this information is in order, the Recreation Coordinator will recommend to the Association that the organization be chartered as an ATRA-sponsored activity.

If the ATRA Activities Council approves the charter, the organization is expected to conform to general policies as outlined below.

Elected activity officers for all clubs/leagues must realize they assume a responsibility to their fellow officers, to their members, to ATRA and the Company. They must give leadership and direction to the adherence of by-laws and ATRA policies.

Section E. <u>Finance</u>. All association funds other than Activity fees shall be deposited with ATRA. ATRA will issue checks for approved expenses.

Collect and auditable bookkeeping records must be maintained at all times with all club/league officers being responsible for accountability.

Each organization must submit activity and financial reports to ATRA each year to qualify for continued sponsorship and financial assistance. Financial reports are compiled into the total ATRA tax exemption non-profit status and are therefore subject to ATRA, Company, State, and Federal audit. They must be completed accurately according to proper bookkeeping procedure, authenticated by signatures of the appropriate club/league officers.

Donations or door prize collections must be receipted by the club/league treasurer or president.

All ATRA approved expenditures must be authorized by action of the organization's executive committee and so stated in minutes. At no time can individuals expend funds without explicit approval from the club's/league's executive committee.

Financial arrangements between ATRA and club/league activities are as follows:

1. To be eligible to apply for financial assistance an activity must have 100% of its members belonging to ATRA.

2. Clubs/leagues are required to submit an itemized budget to ATRA's Recreation Coordinator. The budget must show a basis for financial planning, and reflect:
   a) anticipated membership
   b) anticipated income from all sources
   c) anticipated costs of equipment
   d) anticipated entry fees, ground fees, etc.
   e) an itemized statement of financial affairs for the preceding fiscal year and an estimated budget for the coming year.

3. ATRA shall provide financial support it feels ATRA can afford for items c and d after the remaining information has been provided. The following restrictions shall apply:
   a) All items of a personal nature shall be considered the financial responsibility of the participants.
   b) Capital equipment purchases with ATRA support shall be used exclusively in ATRA-related activities.
   c) No items of an expendable nature shall be purchased with ATRA funds (i.e., refreshments) unless approved in advance by the ATRA Recreation Coordinator.

Section F. _Additional ATRA Assistance_. The preceding services and assistance will be in addition to other services provided by affiliation with ATRA which are as follows:

1. Personal Liability and Property Damage insurance.
2. Trophies, awards, etc.
3. ATRA publicity
   a) bulletin boards
   b) newsletters
4. Reproduction service
5. Meeting rooms and activity facilities
6. Accounting services
7. Checking account

8. Banking of funds
9. Purchase and loan of sports equipment for ATRA activities. A receipt will be required on loan of all equipment; it will be returned immediately upon return of the loaded equipment. If damaged, appropriate assessment will be made to replace it.
10. Liason between management and clubs/leagues.

Each activity receiving assistance shall elect from among its officers a "property custodian" who will report to the Recreation Coordinator of ATRA. The custodian's function is to certify that ATRA support is used properly and that equipment is adequately cared for. All requests must be submitted to the Recreation Coordinator for review. If in order, the request will then be submitted to the ATRA Activities Council for consideration.

Section G. <u>Property</u>. Any major procurement of equipment must be requested in writing and approved by ATRA. Any sale of major ATRA equipment must have ATRA's Board of Director's final approval. Strict accountability will be required for all property. All equipment and funds shall be considered ATRA property. In case of club/league disbandment, all properties and funds will revert to ATRA for deposit or to be placed in trust in the event that the activity should reorganize.

Section H. <u>General Policies</u>. No ATRA club/league shall obligate itself or ATRA to term payment or contracts without first obtaining written approval from ATRA.

Publicity in newsletters, local papers, magazine publications, or spot radio and TV announcements must be approved by ATRA prior to submittal.

Members acting as club/league representatives shall act in accordance with the Articles of Incorporation and By-Laws of ATRA, as amended, and shall not engage in any activities of a political nature, and will not become involved in matters which are recognized functions of Ashland Oil's management.

Each activity must submit, as requested, an activity report to the ARTA Recreation Coordinator. The report shall detail activities completed during the past month, and activities planned for the coming month. Changes to the club's or league's roster must be reported monthly.

Section I. <u>Indemnification</u>. The corporation shall indemnify to the full extent authorized or permitted by the laws of the Commonwealth of Kentucky, any person made, or threatened to be made, a party to an action, suit, or proceeding (whether civil, criminal, administrative, or investigative) by reason of the fact that he/she, his/her testator or intestate is or was a director, officer, or employee of the corporation or serves or served any other enterprise at the request of the corporation.

ARTICLE IX: FISCAL YEAR

Section A.   The fiscal year of the association shall start September 1st and end August 31st.

ARTICLE X: AMENDMENTS

Section A.   These By Laws may be amended by a majority vote of the entire Activities Council and of the Policy Committee, provided notice of the proposed change has been made at a previous regular monthly meeting.

Appendix C:
Sample Copy of a Job Description of a Director
(Lockheed Employees Recreation Club; Burbank, CA)

Executive Director: L.E.R.C.
Reports to: Director of Industrial Relations

Functions:

Administer a program of leisure time activities and services with the objective to develop a healthier, happier, and more motivated employee who will respond with loyalty and a more productive effort.

The services should offer diversification in order to allow more employees the opportunity to enrich their lives in wholesome leisure time activities and to give additional benefits in cost savings programs.

The operation must be well organized with sound business procedures, develop programs to assist with the financial burden and maintain a good image for Lockheed and the L.C.R.C.

Responsibilities:

Administration:

1. Act as liaison between the company and the elected Council Representatives with a constant endeavor to keep the arrangement in a positive benefit to both parties.
2. Hire and develop a professional and dedicated staff that will work as a team for a cost effective operation with respected results.
3. Effect an efficient accounting system including computer operation. The growth of the services offered by the L.C.R.C. has necessitated change as the operation is grossing almost a million dollars a year.
4. Submit a budget request to the company justifying financial support for L.E.R.C. functions. Develop detailed and comprehensive budgets for each of the three divisions of the L.E.R.C., as well as the overall administrative opera-

tion. The four budgets are approved by the L.E.R.C.
Board of directors.

5. Cause proper supervision for the operation, mainte-
    nance and repair for Burbank's five acre park with
    two major structures, the recreation building at
    Palmdale including a fitness room and a newly pur-
    chased 24' x 60' trailer for an L.E.R.C. facility at LAS-
    Ontario.

6. Develop plans for future growth and development of
    additional programs, such as a fitness center at Bur-
    bank and a more diversified and beneficial buying
    service for the employees.

7. Research and purchase of modern equipment to ef-
    fect a more proficient operation, such as a computer
    and computer type cash registers, office equipment,
    transportation and maintenance equipment.

8.   Research and purchase adequate insurance to cover
    L.E.R.C.'s liability, fire, auto, workmen's compensa-
    tion and the pistol and rifle programs.

9.   Supervise the payment of and/or reporting of all
    taxes: property, sales, federal income, state income
    and the California Attorney General's office.

10. Develop and sign contracts for services and facilities
    between agencies, the company, outside organiza-
    tions such as the City of Burbank, schools, local
    colleges and vendors.

11. Invest operating and reserve funds in the highest
    interest earning accounts and assume responsibility
    for all funds and assets.

12. Promote and direct safety in all of the L.E.R.C. opera-
    tions and institute security alarm systems in each
    vulnerable area of the facilities.

13. Take an active role in professional and community
    organizations to further develop a more effective and
    healthful leisure for the employees and cause a re-
    spected image for the profession and for Lockheed.

14. Attend seminars to broaden the educational and ex-
    perience base for the director, the staff, and the total
    operation.

15. Assure a proper legal tax exempt entity by adhering to the Articles of Incorporation, the exemption letter, maintaining a proper set of Bylaws and by submitting all required reports on time.

Programs:
1. Work with and assist in every way possible, the volunteers and elected Council Representatives to effect worthwhile and beneficial programs of service and leisure time activities in three separate Councils divided by geographic areas: Burbank, Palmdale and Ontario.
2. Effect a company wide election for approximately seventy-five Representatives from the three areas and then subsequent election of their officers and committees - (election response is equal to most city elections). A six-person Board of Directors with representation from all three areas is the top covering body of the L.E.R.C.
3. Guide, when possible, a proper organizational operation and wise spending of funds.
4. Be responsible for agenda, minutes of all meetings and the communication to effect the proper operation of the total program.
5. Suggest and develop diversified programs to benefit more retirees, employees and their families. Fitness and health is the most important thrust for future development.
6. Implement additional services that have positive financial benefits for the employees as well as the L.E.R.C. Self-sustaining programming is a major goal.
7. Research and develop reports and supervise vendors who become participants in the "referral" programs or other discount offerings.
8. Be a constant force for more cost savings operation and cost effective programming.
9. Assume responsibility for purchasing, inventory, sales, and accounting for the L.E.R.C. novelty store that is grossing approximately $150,000 in sales per year. (A positive benefit to the company in customer relations, public relations, and a moral factor for the employees).

Appendix D:
Sample Copy of a Job Description of an Intern
(Honeywell Employee Club, Minneapolis, MN)

HONEYWELL
POSITION DESCRIPTION

Title: Recreation Student Intern
Division: Minneapolis Operations
Department: Recreation Services
Date: {update constantly}

MAJOR FUNCTION:
The student intern is responsible for assisting the recreation staff
in the planning, organization, promotion, administration, and
the evaluation of assigned recreational services and activities for
all active employees, retirees, and their families.

The incumbent is accountable for completion of one project
during the internship period in which to apply skills and knowledge of planning, administration, communication, and performance evaluation. The incumbent is also accountable for the
completion of a written project during the internship period in
which to apply skills in business communication.

SOURCE OF SUPERVISION:
Manager of Recreation
Student's College Adviser

WORK PERFORMED:
I. Prepare internship objectives and establish tasks to be com-
pleted during the program.
   A. Prepare intern objectives and timeline that will com-
      ply with college program guidelines and submit to
      the college adviser and intern supervisor.

B.   Complete a midquarter evaluation and submit second half objectives to supervisor and adviser.

C.   Periodically meet with intern supervisor and adviser to discuss progress.

D.   Complete a final evaluation and performance appraisal of intern accomplishments, problems encountered, and recommendations for future intern programs.

II.   Obtain information regarding the structure and function of the employee recreation program and its role in human resources management.

A.   Attend the new employee orientation meeting to gain an understanding of company philosophies and goals.

B.   Study company Policy Manual.

C.   Study Recreation department objectives and policy manual.

D.   Visit other local employee recreation programs.

E.   Read material relevant to trends and changes influencing employee recreation programs.

F.   Schedule appointment with company staffing representative to obtain information about job interview process, resume, and career opportunities.

G.   Schedule appointments with representatives of departments and/or vendors who provide relevant services to the Recreation department.

III.   Assist the Recreation department staff with assigned programs.

A.   Attend any planning committee meetings for events and assist programmers in the implementation of those events.

B.   Assist in procuring materials and services from vendors and supporting company departments.

C.   Attend weekly staff meetings to report progress.

D.   Assist in evaluation employee events and activities.

IV.   Complete one project individually, approved by intern supervisor and college adviser.

A.   Develop objectives.

      B. Establish strategies to meet those objectives.

      C. Assess resources and prepare budget, recruit volunteers, determine facilities, etc.

      D. Prepare promotion and publicity.

      E. Interface with and procure services and required materials from vendors and supporting company departments.

      F. Secure final arrangements. Execute purchase/service requisitions as needed.

      G. Implement and supervise the activity.

      H. Monitor program or event.

      I. Evaluate and prepare written debriefing for intern supervisor.

      J. Maintain an activity file.

V.   Prepare written project as approved by intern supervisor and college adviser.

      A. Assess potential topics.

      B. Gather data and prepare topic outline.

      C. Prepare timeline.

      D. Report progress to supervisor and submit completed project.

VI.   Assist with a divisional project that permits extensive contact with a volunteer committee, as mutually agreed upon by the intern supervisor, intern, and divisional representative.

      A. Clearly establish role and specific assignments for divisional project.

      B. Establish timeline for tasks.

      C. Report progress to intern supervisor.

      D. Participate (on-site) until the successful completion of project/function.

      E. Prepare debriefing for intern supervisor and divisional representative.

VII. Miscellaneous assignments

      A. Assist in answering questions and solving routine problems as needed.

      B. Maintain activity files.

C. Assist with walk-in/mail order discount ticket sales.
D. Interface with professional associations as approved by intern supervisor.
E. Develop word processing, data base, and spreadsheet skills utilizing personal computers provided by the company.

Appendix E:
Sample Copy of an Income and Expenses Statement from an Employee Association (Lewisville Texins Association, Lewisville, TX)

Income Statement
For the Period Ending December 31

Income	
Vending	$51, 064.00
Interest	$7,605.36
Club and Activities	$75,404.27
Club Memberships	$650.00
Rentals	$87.00
Total Income	$134,810.63

Expenses	
Administration Expenses	
Accounting Fees	$2,062.95
Depreciation	$5,697.24
Dues	$265.00
Insurance	$550.00
Office Supplies	$166.11
Supplies	$10.48
Travel and Entertainment	$776.39
	$9,528.17

Clubs and Activities Expenses	
Maintenance and Repairs	$460.90
Office Supplies	$37.40
Outside Services	$18,039.75
Rentals	$36.268.09
Sport Supplies	$3713.15
Tournament and League Fees	$3,782.12
Travel and Entertainment	$3,782.12
Trophies and Awards	$15,108.76

Appendix E (cont.)

Utilities

$1,635.35

<u>81,382.50</u>

Total Expenses

<u>$90,910.67</u>

Net Income

<u>$43,899.96</u>

Answer Sheet

What are recreational activities for Honeywellers all about?

For over 60 years Honeywellers have enjoyed organized social and recreational activities together. In the beginning, activities were coordinated by the Minnregs. Since 1977 they have been coordinated through the Recreation Department. Today, participation has never been higher. Honeywellers continue to enjoy a wide variety of recreational activities and services.

The next step in the growth of the recreational services program is the establishment of the Honeywell Employee Club Minneapolis, Inc. The new Employee Club is for **all** employees/retirees and includes all the activities, sports, special interest groups and discount services previously available through the Recreation Department with exclusive benefits offered to its members only.

Why the change? Is it necessary?

A substantial number of Honeywellers and their families participate on an annual basis. The rate of increase between 1983-85 was 79%. While Honeywell has agreed to provide a significant portion of the funding for the program, the level of funding must be limited. A cost sharing system administered through an employee directed club will help to maintain program levels for the future and fund new opportunities that will positively affect employees.

How is this different from what we had with the Honeywell Recreation program?

The Employee Club is different from the old Honeywell Recreation program in two major ways:

The Employee Club is led by a Board of Directors composed of employees. In the old recreation program no employee board was authorized to make decisions concerning employee interests. Now employees are elected to serve on the Board of Directors for a two year term. The Board is authorized to set policy and determine to what extent recreational activities, events and employee services will be financially sponsored. This increased employee involvement will help promote employee participation.

Second, a nominal annual membership fee is charged to help defray costs of Employee Club activities and services. While there was no membership fee for the old recreation program, the program relied almost exclusively on funding from Honeywell. As mentioned, this was not feasible to meet the rapid increase in employee participation. By charging a membership fee, activities and services can be maintained and member dues used to develop desired programs.

How is the new Employee Club organized?

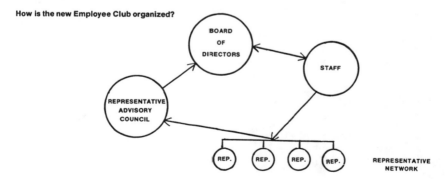

Board of Directors

The Board of Directors is composed of salaried and hourly employees, seven members total. The Board of Directors are responsible to members of the Employee Club and on a general basis to Honeywell.

Representative Advisory Council

A cross section of employees are appointed to serve as the advisory body to the Board of Directors. This group is called the Representative Advisory Council (RAC). The RAC meets 2-3 times a year with the Board to confer and recommend ways and means to improve the Employee Club.

Employee Club Reps

Interested employees serve as Employee Club Representatives (Reps) for their department or area. They circulate information about Club activities, discounts, Board of Directors decisions, provide information to new employees and give them the opportunity to become a member of the Employee club.

Employee Club Staff

A full-time staff, formerly known as the Recreation Department, is responsible to administer the day to day affairs of the Employee Club. The Employee Club office is located in Honeywell Plaza. The staff also handles all-Minneapolis special events and discount programs which provide reduced pricing for vacations, family amusement, theater, professional sports and other merchandise and service programs. All events and programs are announced through the Employee Club newsletter, company bulletin boards and the Honeywell Circulator.

What happens to the Minnregs, Honeybelles and other special interest recreational groups under the Employee Club plan?

There is no change under the new Employee Club. All special interest groups under the old Honeywell Recreation department, including the Minnregs and Honeybelles, will continue to promote activities. There will be individual fees to join each of these groups.

What happens to the Minnregs, Honeybelles and other special interest recreational groups under the Employee Club plan?

There is no change under the new Employee Club. All special interest groups under the old Honeywell Recreation department, including the Minnregs and Honeybelles, will continue to promote activities. There will be individual fees to join each of these groups.

Will I be able to use the program if I'm not a member?

The Board of Directors will permit employees who are not members to participate in certain Employee Club programs, however, nonmembers will pay higher fees for individual activities. Please refer to the chart below for specific member and nonmember benefits.

A.   Exclusive Member Benefits

Discount Tickets
National Theme Park Discounts
Employee Club Family Day*
Employee Club Special Events*
Free Wholesale Club Membership
Employee Club Newsletter

*At a reduced rate

B. Non-member Rates

Non-members may participate in these programs at an additional fee per event or where a guest policy is designated such as:

Sport teams, leagues and tournaments
Special Interest Clubs
All-Minneapolis Events
Education Programs

C. For members and nonmembers

Discount List
Discount application programs
Employee Store/Clothing programs
Equipment Rental programs
Film Processing programs
Holiday Greeting Cards and Stationery
MTC Bus Card Service

Why should I become a member of the Employee Club? What are the advantages to me?

The Board of Directors encourage you to join the Employee Club for several reasons:

First, your membership is the easiest and most economical way to participate in the program. Remember, only members may purchase discount tickets. Nonmembers may participate in selected programs but will be charged an additional fee per event.

Second, your membership allows you to influence your fellow employees by serving on the Board of Directors. This is important when decisions are made to fund activities you enjoy, or expand services.

Third, employee club members vote on issues that the Board of Directors determine have far reaching impact on the Employee Club. This means that more employees will have involvement in the success of recreational activities and services than ever before.

In addition, membership funds are carefully managed. As mentioned, member dues are combined with financial support from Honeywell and other activity fees to help support the Employee Club. This means the Board of Directors can use member dues to fund Club programs that would not be possible using only financial support from Honeywell.

How will my $5.00 be used?

The Board of Directors determines how membership dollars are spent. The Board could choose to sponsor a major family event or several events. The Board could also choose to offset operating costs for the Club or to invest in a major project instead. The Board's options are limited only by its creativity.

When can I join the Employee Club?

You may join the Employee Club anytime of the year; however, you are encouraged to join early so you enjoy club benefits for the entire year. The cost to join the Club is the same throughout the year and entitles you to Employee Club benefits through December 31 of the year you join.

Can I get a refund on my membership? Is the membership transferable?

Membership fees will not be refunded. If you choose to withdraw from the Club before the next payroll deduction, you must complete a Membership Withdrawal Form and turn in your Employee Club card before December 31. Employee Club memberships are not transferable.

Employee Club By-laws

A copy of the Employee Club By-laws may be obtained by calling 870-6666. For more information or questions call 870-6666.

Appendix G
Sample Copy of a Liability Waiver

All activity members are required to read, agree to, and sign this waiver before entering a recreational activity.

I, _____, (please print) do for myself, my heirs, executors, administrators, and assignees waive, release, and forever discharge any and all rights and claims which may have or which hereafter accrue to me against the {your employee association's full name goes here}, including officers, employees, and instructors. I assume all risks associated with this event including, but not limited to, falls, contact with other participants, the effects of the weather, including cold, heat and humidity, traffic and conditions of my assigned playing area, including roads during races, all such risks being known and appreciated by me. Having read this waiver and knowing these facts and in consideration of my participation, I for myself and anyone entitled to act on my behalf, waive, release my instructors, captains, and the {your employee association's full name goes here} along with any sponsors, their representatives and successors from all claims or liabilities of any kind arising out of my participation in this event.

Entry must be signed by participant.

_____   _____

Signature   Date

1987
ANNUAL REPORT

HONEYWELL EMPLOYEE CLUB
OF MINNEAPOLIS, INC.

1987
ANNUAL REPORT OF THE
HONEYWELL EMPLOYEE CLUB OF MINNEAPOLIS, INC.

1987 was the first year of operation for the Honeywell Employee Club of Minneapolis, Inc. (EC). This was a year of transition as we got use to the Recreation Services Department becoming the Employee Club. As your employee Board of Directors, we were excited and enthusiastic about our role to set the direction for the new organization. The staff was also busy reorganizing and changing the Recreation Services Department office in order to coordinate the business affairs of the new club.

We began the year by announcing that current employees and retired employees of Honeywell in the local area could join the EC by enrolling for an annual fee of $5.00. We must admit that we were not sure how employees would respond, but as it turned out, employee response was greater than anyone expected. The months following the announcement were busy ones as over 10,000 Honeywellers enrolled in the EC.

This annual report will recap the achievements of the EC in 1987, but more importantly it will describe an organization that came into its own the very first year because of employee support. While it was true that Honeywell provided a substantial portion of the income for the new club, the EC was successful because of employee interest and involvement. We enjoyed serving you and look foward to working together in our second year of operation.

The EC Board of Directors and Staff

SPECIAL EVENTS
Two major special events were sponsored. On June 21st Father's Day, 2,924 members and their guests enjoyed a day at Valleyfair. While the weather was hot, many dads and grandfathers enjoyed ice cream, pop and a chance to cool-off on Valleyfair's "Flume" and "Thunder Canyon" rides with their families. On August 22nd, 4,139 members and guests enjoyed a day at the Renaissance Festival. In the days which preceeded the Renaissance Festival Day, members of the cast could be seen around many Honeywell facilities entertaining employees and promoting the event.

DISCOUNTS
The Wholesale Club program was introduced with the start of the EC. Members were able to purchase many major brand items at wholesale cost plus 5%. Five (5) additional ticket outlets and four (4) additional film processing units were installed in company locations in 1987 as well. The Employee Club Discount Directory was updated and issued twice in 1987. Over 200 businesses offered reduced price arrangements on a variety of major purchase items from automobiles to home furnishings as well as discounts on health club memberships and personal services.

PARTICIPANT SPORTS
Program expansion was evident as EC members participated in leagues and tournaments in 16 communities, as compared to 13 in 1986. Program expansion was also possible in 1987 as new facilities became available during the year. For the first time in four years, repairs of the Golden Valley plant soccer fields were complete and our soccer teams were able to utilize both playing fields.

CLUBS/ORGANIZATIONS
The Concert Band celebrated its 50th anniversary in 1987 and performed at 18 functions associated with Honeywell or the community. This same community spirit was evident in other groups including, but not limited to the Minnregs, Honeybelles, Chorus and Amateur Radio Club. Nineteen (19) special interest clubs and organizations were supported.

OPERATIONAL
A bi-monthly EC newsletter was started and mailed to each members home and featured extended coverage of EC developments. The newsletter and Discount Directory were published entirely by the EC staff utilizing the latest in desk-top publishing technology. A fully integrated computerized accounting system was installed and the new club began to utilize its own separate accrual accounting system for general ledger, order entry, purchase order and inventory functions. Member records were also computerized to send deductions to payroll, produce labels, member cards, rosters and reports.

PARTICIPATION

LEADERSHIP

BOARD OF DIRECTORS

Dorothy Guanella, President	CBG
Jim Winkler, Vice President	MAvD
Tom Nelson, Treasurer	DSD
Bob Crunstedt, Secretary	Staff
Myron Amundson, Director	Corp.
Phil Smith, Director	R&BCG

REPRESENTATIVES COUNCIL

John Catherson	DSD
Jim Farley	CBG
Mary Joseph	CBG
Charles King	USD
Chuck Koch	DSD
Fred Kohanek	R&BCG
Donna Manders	ASTD
Derrell Martin	SRC
Donna Miller	SSED
Bob Pierson	MAvD
Jane Sisson	Corporate
Geno Wincek	USD
Kenny Wood	R&BCG

STAFF

Bob Crunstedt, Manager
Debbie Dahl, Special Services Administrator
Mark Saari, Sports & Recreation Admin.
Karen Moore, Member Records Assistant
Sandy Schmidt, Accounting Assistant
Linda Norberg, Secretary/receptionist

NOMINATING COMMITTEE

Tom Nelson, Chairman	DSD
John Catherson	DSD
Wendy Halverson	SRC
Derrell Martin	SRC
Diane Signorelli	MAvD
Phil Smith	R&BCG

DISCOUNT DIRECTORY COMMITTEE

Dorothy Guanella	CBG
Phil Smith	R&BCG
Jim Winkler	MAvD
Bob Crunstedt	Staff

MEMBERSHIP

STATISTICS

MONTH	ENROLLMENT
January	4,434
February	3,226
March	629
April	764
May	249
June	303
July	251
August	274
September	280
October	199
November	106
December	91
TOTAL	11,429

SERVICES

DISCOUNT TICKET SALES

General Cinema	8,747
United Artists	8,435
Cinemaland	625
Engler	130
Valleyfair	4,285
Renaissance Festival	2,188
Minnesota State Fair	3,807
Minnesota Zoo	419
Minnesota State Fishing licenses	514
Canterbury Downs	2,036
Chanhassen	798
MN Science Museum	264
Ordway Theater Presents	61
Minnesota Twins	1,610
Minnesota North Stars	552
Minnesota Strikers	163
U of M football	112
U of M hockey	76
MTC 10-Ride Cards	202
MTC All-You-Can-Ride Cards	551
Disney's Magic Kingdom on Ice	377
Zuhrah Shrine Circus	127
Osman Shrine Circus	320
Sesame Street Live	496
NBA Exhibition Basketball	67
Pavarotti - Met Center	64
Masters of the Universe - Met Center	60
Ice Capades	60
Wisconsin Dells Funbooks	216
Team USA Hockey	268
Twin Cities Fine Dining Books	34
TOTAL	37,664

PROGRAMS

SPORTS

Basketball	101
Bowling	924
Broomball	365
Football	92
Golf	670
Soccer	463
Softball	1,230
Tennis Ladder	94
Volleyball	732
Bowling Tournaments	776
Golf Tournaments	340
Softball Tournaments	520
TOTAL	6,307

CLUBS/ORGANIZATIONS

Amatuer Radio	60
Bridge	51
Bicycling	65
Chorus	28
Concert Band	41
Flying Club	120
Hockey	41
Honeybelles	2,200
Minnregs	4,200
Motorcycle	28
Pistol	18
Running Club	425
Skeet Shooting	82
Ski Club	152
Square Dance	159
Trap Shooting	102
Travel Club	212
Tennis (summer)	56
Tennis (winter)	49
TOTAL	8,089

SPECIAL EVENTS

Renaissance Festival Day	4,139
Valleyfair Day	2,924
Retiree Social	1,600
Holiday Craft Fair	80
TOTAL	8,743

1987 BALANCE SHEET AND INCOME STATEMENT OF THE HONEYWELL EMPLOYEE CLUB OF MINNEAPOLIS, INC.*

BALANCE SHEET   December 31, 1987

ASSETS

Cash (First Minnesota Bank & Honeywell Employee Credit Union)	$ 14,271
Accounts Receivable	140
Inventory	8,429
Total Assets	**$ 22,840**

LIABILITIES & FUND BALANCE

Accounts Payable	$ 35,747
Deferred Revenue:	
Pre-paid memberships	2,160
Gift Certificates	800
Liabilities	38,707
Fund Balance	(15,867)
Total Liabilities & Fund Balance	**$ 22,840**

INCOME STATEMENT   Year Ended December 31, 1987

REVENUES

Contributions	
Honeywell Inc.	$300,000
Other	2,792
Memberships	53,663
Sales	319,895
Program fees	40,553
Interest income	9,583
Total Revenues	**726,486**

EXPENSES

Cost of sales	263,633
Program expense	197,318
General & Administrative:	
Salaries & benefits	172,882
Employee activities	2,343
Travel & meetings	4,891
Stationery & printing	7,434
Postage	4,754
Telephone	4,296
General supplies	6,757
Purchased services	13,618
Education & training	2,805
Advertising	25,280
Equipment Rental	2,403
Office Rental	24,414
Data processing	5,411
Other	4,114
Total Expenses	**742,353**
Excess of expenses over revenues	(15,867)
Fund balance year begin	-
Fund balance year end	**(15,867)**

*This financial summary is included in the audited financial statements released by the independent auditors of Stirtz, Bernards and Co., Certified Public Accountants

Appendix I:
Sample Copy of Club Guidelines
(Chemical Abstracts Service, Columbus, OH)

1. To form a club, interest from CAS staff must be demonstrated. If it seems that number of CAS staff are interested in forming a club, they may submit an itemized request to CARSA for funds with which to start the club. Such a request would then be discussed and voted upon by all CASRA board members. This may occur at any time of the year and is distinct from annual funding.

2. Each club sponsored by CASRA must draft a set of guidelines or bylaws which would describe the purpose of the club, election, and length of term of officers, annual fees, etc. Such guidelines and bylaws of established clubs are available for use as examples.

3. For a club to receive annual funding, the following criteria must be met:

    a) Annual membership fees must be collected from all members of the club in an effort to fund club activities and equipment needs. Annual membership fees are to be set by each club.

    b) The club must have an active membership of at least 10 CAS employees. A list of active club members must be kept on file with the CASRA liaison and updated twice yearly (Jan-Feb and July-Aug)). Active members are those members who continue to pay their annual membership fees.

    c) As a minimum, each club should have an elected President and Treasurer. The names of these individuals should be noted on the active membership list.

    d) Twice yearly, prior to submitting an updated active membership list to the CARSA liaison, the club must, via the *NewsCASter* or other suitable media, solicit new members. The announcement of the time, place, and content of the club's regular meetings should be included in the solicitation, along with an invitation for potential new members to attend the next scheduled meeting.

   e)   The preceding criteria must be followed in order for a club to receive annual funding. Failure to do so will result in no annual funding being provided to a club during that year. In the case where funds have already been distributed, the club will be denied annual funding for the following year.

4.   Equipment purchased with CASRA funds remains the property of CASRA and must be labelled and signed out from the designated CASRA board member. The individual and the club are responsible for the equipment which they sign out. Examples of such equipment would include, but is not limited to, game boards, uniforms, and books. Magazines, T-shirts, etc., would not remain property of CASRA.

5.   A club is eligible for annual funding in the CASRA fiscal year following the year in which the club is formed. To receive annual funding, an itemized list of proposed expenditures must be prepared and submitted when CASRA announces that such budgets are due to be submitted.

6.   The maximum amount CASRA will provide to any single club is set at $50.

Appendix J:
Sample Checklist for Operating an
Employee Recreation Activity
(Chemical Abstracts Service, Columbus, OH)

CHECKLIST FOR OPERATING AN EMPLOYEE ACTIVITY

1. Make all outside contacts well in advance of the activity (arrange for speakers, refreshments, music, etc.). Terms of a contract, prices, delivery dates, etc., must be by written agreement. Verbal agreements are subject to change, often to our disadvantage. Any contracts should only be signed by Jenny Wood or Wayne Mills.

2. Reserve any (CAS) rooms early. Check with the lobby receptionist. To reserve the cafeteria, contact Pete Johnson.

3. Notify Frank Healy three weeks in advance about any activity being held at CAS. He needs this information for security purposes.

4. Submit a publicity article to Polly Dougherty for publication in the *NewsCASter*. Follow the production schedule for every two week deadlines. See examples of any clip-and-send. The *NewsCASter* can be a valuable advertising tool. Include articles up to two months before the event.

5. Order posters and tickets at least six weeks prior to the activity. Contact Jenny Wood for the proper forms to have them done in-house. Or try it yourself - you might be pleasantly surprised. If planning to use the showcase, contact J. Wood.

6. Be prepared to talk about the event preparation at the CASRA meeting prior to the actual event. If you need people to help with ticket sales, have sign-up sheets ready.

7. Put up posters to publicize the activity two to three weeks ahead of time. J. Wood can OK them. Contact Emett Crawley if photos are to be taken during the activity. He might be able to have someone take photos. He can almost always loan you a camera and film to take photos yourself!

8. Work orders need to be submitted 4-5 days in advance.

9. Keep all receipts for items purchased and turn these in to J. Wood for reimbursement.

10. Submit article to *NewsCASter* for summary of the event.
11. Write a post-activity report and submit this to the president immediately after the event. Additional correspondence (ex. memos, articles, art, clip-and-send) should be kept with the post-activity report.
12. Give a final report on the event at the CASRA meeting following the event.
13. A file of past CASRA activities is located in Jenny's office. You may wish to refer to these for ideas or past contacts. Keep good records so that the next year CASRA representative who takes on a particular event will be able to refer to these records also. When you are finished with past CASRA records, please return them.